IT CAN
BE DONE

IT CAN BE DONE

AN ORDINARY MAN'S EXTRAORDINARY SUCCESS

by Chick Stewart

with Michele Carter

ENJOY MY STORY.
CHICK STEWART

HARBOUR PUBLISHING

www.harbourpublishing.com

Harbour Publishing Co. Ltd.
P.O. Box 219, Madeira Park, BC, VON 2H0
www.harbourpublishing.com

Wood texture on page viii and endsheets by texturefabrik.com
All photographs not otherwise credited are from the author's collection.
Dust jacket and text design by Setareh Ashrafologhalai
Printed and bound in Canada
Printed on 10% PCW and FSC certified paper

Harbour Publishing acknowledges the support of the Canada Council for the Arts, which last year invested $153 million to bring the arts to Canadians throughout the country. We also gratefully acknowledge financial support from the Government of Canada through the Canada Book Fund and from the Province of British Columbia through the BC Arts Council and the Book Publishing Tax Credit.

LIBRARY AND ARCHIVES CANADA CATALOGUING IN PUBLICATION

Stewart, Chick, 1928-, author
    It can be done : an ordinary man's extraordinary success /
by Chick Stewart with Michele Carter.

ISBN 978-1-55017-800-5 (hardcover)

    1. Stewart, Chick, 1928-. 2. S & R Sawmills (Surrey, B.C.)—History. 3. Sawmills—British Columbia—Surrey—History. 4. Businessmen—British Columbia—Biography. I. Carter, Michele, 1950-, author II. Title.

HD9764.C32S738 2017   338.4'76740092   C2017-902696-8

In loving memory of my wife Marilyn
who still inspires me every day. I miss you dearly.

CONTENTS

SELF-EMPLOYED
CHILDHOOD

C ALL ME CHICK. I've been called Chick since I was six years old. If you call me Donald, I'll know you don't know me. In this story, I'll tell you how my life unfolded over the last eight decades—how I got that nickname; how I met and married the most beautiful girl in the world; and how I came to own and operate S & R Sawmills in Surrey, British Columbia. By the end of this book, I'm pretty sure you'll know me well enough to call me Chick.

My story starts in a small district of Winnipeg in Manitoba. Our family's chicken farm in Fort Rouge bordered a cemetery, where a couple of gravediggers encouraged me to do something unusual. Don't be afraid, they said. I learned early on that when an opportunity crosses your path, take it, even though it might make you uneasy at first.

I was probably six or seven the first time I went over the wall into the cemetery. Our large chicken coop was against the cemetery fence on Arnold Avenue, so the chickens would sometimes walk along the shingles at night and then flutter over the fence into a tree. My brother Dave was a year older (born July 6, 1927), and he showed me how to get over the fence to bring the chickens back.

I climbed over and had to step on a gravestone before jumping down to the ground. The chickens were sitting up in the branches, so we'd shoo them back to our yard. I wasn't afraid of going in the cemetery because I'd grown up beside it. The whole family was used to it. My mom always knew when someone had passed away because suddenly a bouquet of fresh flowers appeared on the kitchen table. In the evenings, we'd dare each other to run around the cemetery in the dark. If other kids came along, we'd dare them too; we knew they were scared when they'd bail out and take the shortcut around the chapel.

The chapel had a morgue for the bodies that couldn't be buried until summer. During the winter the ground was frozen, so the gravediggers weren't around. People were still dying, though, and the bodies went to the morgue after the funeral services to wait their turn. The spring thaw was still too soon to dig, so a number of bodies could be waiting a long time for a grave.

In the winter, everything was quiet at the cemetery, but summer was more interesting because new graves were getting dug. The gravediggers were two middle-aged men named Mike and Pete. They'd get their shovels and dig holes eight feet long and five feet deep. It was hard work in the heat, so we'd climb the fence to bring them water. We'd see several holes already prepared for the bodies and listen for the scraping of the shovels to find where they were working that day. We'd talk to them and get to know them.

One day, the men brought out a few coffins and put them beside the holes. The coffins could be from services that took place months before and the bodies inside could be decomposing in different stages. My brother and I brought over water; we wanted to watch the men bury the coffins. We were surprised when Mike lifted the front end of a coffin and said, "Let's see who we got here." Dave and

I stepped a little closer. We were curious; we'd never seen a dead person and thought the body would look normal, but we still kept back until Mike waved us over. Do we look or do we turn around and go home? A bird in the hand, I figured, and so did Dave. We leaned in and saw a grisly sight. We just about jumped out of our pants. This one had been dead awhile and looked pretty bad. I shuddered at first and looked away, but then took a second look. I remember seeing the man's pocket watch and chain. He had on a tie and suit, dressed in his best. We knew the gravediggers were trying to scare us, but we didn't want to back down. We were like most kids in that we liked to be scared, and a cemetery could be a scary place for people.

Not everybody could afford to be buried in a casket. After many years of burials in a small cemetery, I guess the remains could surface unexpectedly. My younger brother Herb remembers finding a skull and at first thinking it was a soccer ball.

My mom had an experience with the graveyard that got her into trouble. Her clothesline went from our back porch over the chicken barn to a tree by the graveyard. She put out the sheets on the line on a windy day to get them dry. She didn't know it, but one of the sheets got caught in the pulley by the tree. A lady was walking by and screamed. She told a neighbour she'd seen a ghost, so the neighbour called the police. The policeman gave Mom heck because her sheets were causing a disturbance. Dave and I had to climb up the tree and pull the sheet out of the pulley. We saw this as an adventure and looked forward to new ones.

Our twelve-acre farm had a couple of hundred chickens that always needed new additions to the flock. Whenever my dad travelled to the outdoor market for more chickens, he'd take one of us on the three-hour walk into the north end of Winnipeg. We'd walk past Hudson's Bay and Eaton's department stores at Portage and

Main on our way to the market to buy white leghorns—they were good layers. He'd look at how the white comb was lying over the top of their heads to find a healthy one. We'd get Barred Rock chickens and the Wyandotte for eating. They're both stout breeds, meatier than the streamlined leghorns. We'd buy fresh chickens and carry them home in a wire basket. We were more interested in selling eggs than chickens, but sometimes sold a couple of leghorns. Occasionally, we'd peddle eggs around town, but our customers mostly came to the farm.

I got my nickname because of the farm. When Dave went into Grade 2, Mom said, "Take Donald and get him started." I was going to Grade 1.

In the schoolyard, some older boys asked, "Who's this guy?" I looked at Dave to see what I should do, and the boys said, "What's your name?"

Dave nudged me to answer, so I said, "I'm Donald Stewart."

"What's your dad do?"

"My dad's got a chicken farm. He's got a bunch of chickens."

Dave led me into the classroom and then walked with me to school the next day. When we got to the schoolyard, the same boys called out, "Chick, chick, chick—here comes Chick."

After that, everybody called me Chick, and it's my name today. If I ever get a letter addressed to Donald Stewart, I'm pretty sure it's not from anyone I know. Most people who know me call me Chick.

Our family soon grew to six kids. By 1934, I had four brothers and one sister. Dave came first in 1927; I was born September 11, 1928; Denny was born in 1929, Sam in 1930, Margaret in 1933, and Herbert in 1934. As the babies kept coming, Dave and I were wondering when they were going to stop. My sister Marg was number five, and my mom was pleased she finally had a girl. My dad didn't

My parents borrowed this horse and buggy for our photograph. Dave's holding the reins; Denny, me and Margaret are behind him.

drive, so the insurance man would drive my mom and the babies home from the hospital. We had two hospitals at the end of Morley Avenue. I had to go to the hospital when I had the measles, and my brothers went there to get their tonsils out. I found out they were going to get ice cream, and I cried when I couldn't go with them because the doctor said my tonsils were okay.

We lived in a good-sized house with an upstairs. We'd sleep two in a bed, and in winter we'd keep our boots in the room. We'd never leave them outside or they'd freeze. In that cold weather, we'd wear long johns under our clothes and as our pyjamas. Our pot-bellied stove in the kitchen burned wood, but sometimes we'd stoke it pretty heavy with coal because coal lasted longer, especially over-night. The heat would rise upstairs and keep us fairly warm, but it'd be chilly in the morning. We had a well for water and tap water in the faucet to do our dishes, and a bathroom in the house, but we had an outhouse as well.

We never felt poor. Our chicken farm made good money, and we had a big garden, but we didn't have a lot of spare money. We didn't have hardships, but it was tough going sometimes because we had uncertain finances. The relief places in churches would give us a bag of clothes, and we'd get hampers from the Salvation Army at Christmas. We'd wear hand-me-downs, and sometimes people would donate boots for us. We mostly wore leather moccasins with laces in winter, and went barefoot in summer. Everybody went barefoot. We didn't see anyone wearing fancy sandals in those days. Moccasins and leather boots were enough for us in winter.

We were always too busy to spend time thinking about our situation. We had plenty to do every season. We would work in the potato garden, either pulling up potatoes or making rows to plant them. My brothers and I would plant a row of potatoes before school, and after school we'd spend an hour in the garden weeding and keeping things clean. People would come round to buy our potatoes.

We had a cow and I knew how to milk her. Our cats were barn cats and they ate mice, but I'd sometimes shoot milk into their mouths for an extra treat. We also bought buttermilk from a guy who'd bring it in a big tin container. We'd feed the chickens with it, and we drank it too. I kept a few rabbits that liked to dig holes into the chicken pen, where they'd run around. They were sort of wild. We raised them and ate some.

After doing our chores, we had plenty of time to play. We'd go over to River Park to fish. We'd go to the butcher before we went fishing and get a small piece of meat to put on a pin that we used as a hook, and we took a little fishing line to catch crabs. We wouldn't eat them or anything, but it was fun. We never caught any fish in

Winnipeg, and when we went crab fishing we always put the crabs back in the water.

The fire hall was on the corner of Arnold and Osborne, about half a city block away. We would go down there to talk to the firemen and watch them play horseshoes. They let us play too. It was always exciting when the truck would go down the street with its siren blaring. We were never allowed on the truck.

The bakery and milk trucks went up and down the avenues, their horses pulling them. Along the avenues, the houses were side by side, giving the deliverymen more stops on their route. The horses knew the route, so the deliveryman would drop the reins and let his horse saunter along until it stopped at the next customer's house. He'd put an iron poker in the ground and tie the reins to it so the horse wouldn't go to the next customer on its own. We used to watch to see if the horse really knew where to stop, and every time, it would slow right down and come to a halt in front of the right house.

We liked to do odd jobs for a few cents. I'd shovel a sidewalk for ten cents and then go to a movie in Winnipeg. If Dave and I saw a truck come by to dump a pile of wood or a load of coal out back of our neighbours, we'd knock on the door and ask if we could stack the wood in a pile or throw their coal down the chute. We'd get a nickel or a dime.

We weren't afraid of work. A movie was ten cents, so if we wanted to go, we had to earn the cost. We could see other people were making money and getting things they wanted, so we'd watch how they did it and give it a try. We never missed an opportunity. If we didn't earn money, we couldn't have fun.

Once a week, every Saturday morning, I looked after an elderly lady not far from our house. She was bedridden, so I'd go over and help her. I had three jobs to do for her—I had to take the ashes out of the stove, put them in a metal container and dump them in a corner of her yard; I had to clean the stove; and I had to wash her poodle. I'd put a big pail of water into the tub and tell the dog to get in. He was smart and stepped into the tub and just stood there while I soaped and scrubbed him down. After I dried him off, he'd run around the house, probably glad to be smelling good. The lady had a box of chocolates beside her bed, and she'd give me a chocolate and twenty-five cents. I did that for a few years. I had my nickels and dimes hidden in a tin box in the basement behind a loose brick in the wall. I think my brothers knew it was there, but they never took any of the money.

My parents didn't go to church, but we kids would walk a couple of miles to Sunday school at St. Alban's Anglican Church in Fort Rouge. We went every Sunday and would sing hymns. We went to a church camp in Winnipeg for a week one summer called Manitou. We played baseball there and swam. Our games and swimming were always supervised to keep us safe. We never went away for summer holidays or anything like that. Sometimes we'd play with other boys. We'd walk to River Park amusement grounds, two miles away, when they were running the roller coaster, and we'd watch the guy testing it and ask if we could get on. He let us ride it for free. The front car was the best! I had a Chinese friend named Allan Mah, and his father owned a fish and chip store. After school, we'd go in there and get a little box of french fries for free.

On Halloween we went trick or treating. We never wore fancy costumes, but we'd make cardboard masks and attach an elastic to them. We might wear old toques on our heads too. We'd carry pillowcases and knock on our neighbours' door to get the usual apples and potatoes and a bit of candy. My dad would be waiting at home with a big galvanized bucket for the fruit and vegetables, and he'd tell us to go back out and get more. This way he built up a small stockpile for winter.

We'd get up to a little mischief, but never enough to get a cuff across the ears. We'd tie a black thread to a front door knocker and from our hiding place we'd pull on the thread. Someone would answer the door but we'd be gone. We'd wander in the back alleys and sneak up to an empty outhouse and push it a few feet farther away from the house. All in good fun.

A mom-and-pop store on the corner was called Whitlock's. They'd sell hot cross buns for half price when they got stale, so we'd take a few cents and buy a couple. One time Dave and I played a trick on our little brother Herb. In those days, pennies were a lot bigger than they are today, so we got some silver paper and wrapped the penny to look like a quarter and sent Herb to the store. We wanted him to buy BB Bats toffee suckers that cost two cents. The man at the store knew the quarter was a fake, and he knew us, so he sent Herb home empty-handed.

We didn't have much time to tease our younger brothers, but I do remember putting Herb off eating a piece of raisin pie. Mom had just brought the pie out of the oven, and Herb was digging into it when I came into the kitchen. I leaned over him and said, "Do you know what I put in that pie?" I opened my hand and showed him a few black rabbit pellets. He never ate another piece of raisin pie after that. We still laugh about it.

Whenever our parents went away from the house, the older boys were supposed to babysit. When Herb was about four years old, he discovered he was alone in the house and got scared, so he walked over to Whitlock's to wait until someone came home. He spent two hours over there watching for a light to come on in the house. As soon as he saw a light, he ran home. Everybody had thought the other brother was watching him. We didn't make that mistake too often.

I never played sports. I'd kick a soccer ball around on the playground in good weather but never joined an organized team. In winter, we'd build a snow rink in the backyard and use a hockey stick and a tennis ball. We'd get newspapers and tie them together and put them in our long socks to act as shin pads. All of our noses would end up pure white from the cold, but we were having fun.

Our parents were kind to each other and us. Dad was born in Banbridge in County Down in the north of Ireland in 1899. He met Mom at Eaton's in Winnipeg when he worked there as a caretaker. He'd clean the building after everyone went home. My mom was born in Aldershot in the south of England in 1904, and she came to Vancouver to stay with her brother Roy. My parents were married in Winnipeg on September 4, 1926.

Dad worked hard. He was a quiet man who avoided company. If people came over to visit, he'd leave the room and pretend he had chores in the chicken barn. He never swore or drank, but he was strict and I seldom heard him laugh; he was a serious man. He didn't drive a car but rode a bike instead. He was a good crib player, and I always enjoyed playing cards with him. We'd help him with the chores, sawing wood with crosscuts or hoeing the garden, that kind of thing. We were a hard-working family.

My mom was more outgoing than my dad. She was happy-go-lucky and loved to laugh. She liked to meet and chat with people; I was more like her. When visitors came by, Mom did all the talking because Dad had slipped out the back to avoid the ladies.

Mom had a group of ladies that met to play whist. They would come over to our place, and she would make tea and play cards. When they finished drinking their tea, the ladies would say, "Flo, can you read our tea leaves?" She would tell them to turn their cup upside down in the saucer and then make a silent wish while they turned it three times. They'd hand her the cup and she'd study the way the leaves were lining the cup. She'd shake her head and say, "Doesn't look like you're going to get your wish," or she'd smile and nod and say, "I'm not sure if this was your wish, but I see you're going on a trip. You might be crossing water." It was mostly just for fun. I don't think anyone believed her predictions.

We would ask her to read our tea leaves too. We'd turn the cup three times and make a wish and then hand her the cup. She'd look very serious and say, "Somebody at school is thinking about you. I can see the initial R." Or she'd tell us we were going to the movies to see a little girl perform. We sometimes went to the movies on Saturday night to see a Shirley Temple movie. Mom always said pretty much the same thing, but we'd have a good laugh with her.

Whenever Mom went downtown to grocery shop, we'd wait for her at home. We knew what streetcar she'd be taking back, so we'd make sure we were standing at the stop on Osborne Street to help her carry the bags. She'd always give us a bag of treats as our reward for babysitting the younger children. We didn't do much, but we made sure they didn't get into trouble in the house.

My mom liked to help others. She'd go to an old person's house and clean it for a few cents. They might have some hand-me-downs for us kids, like the boots or moccasins. We were happy to wear them. My mom was also a good seamstress. She made sheets out of the flour sacks. She'd spread them out and iron them, but she never intentionally tried to make them look like ghosts!

She taught us to have good manners. We all sat at the kitchen table for dinner and filled up on her good food. We never got any meat until we ate all our vegetables. If my brother left anything on his plate, I'd eat it. Nothing was wasted. Dad would say, "I think you guys are seeing who can eat the most." Who could blame us? Mom was a good cook, and she baked quite a bit. She made rice puddings, french fries, bread—all good food. She'd get meat from the butcher, so we had enough to eat. She taught us to always say "excuse me" before we left the table. We also helped set the table and shared

The youngest kids with Mom: Herb, Margaret, Florence and Sam.

the after-dinner chores—one was a cleaner off-er, which meant he cleaned the plates, one was a washer, another a dryer, and another a put-'em-away-er. We tried to take turns. We all got along well, and there wasn't too much grumping.

The only accident I remember from around this time happened to my brother Herb. He was in Grade 1 and was walking home in the winter along a frozen path. The snow was about a foot high and the temperature had gone down to twenty below. He was kicking a soup can along the hard ground when his boot got caught in the ice and he fell head first. He put out his hand to break his fall and it landed on a big piece of broken glass. It started bleeding, so he closed up the gash and held onto it while he ran the few blocks home. My dad took him to the hospital, and the doctor gave him fourteen stitches on the palm of his left hand. The doctor told my dad that if it hadn't been so cold, my brother could have bled to death. The freezing cold had stopped the blood flow. It just gelled until he got to the hospital. Winter saved him.

In winter, the Red River froze about three feet thick. People could drive their cars across it and then spin around on the ice for something exciting to do.

We kept warm and dry through the winter. We had the pot-bellied stove and a big furnace that also took firewood or coals. When we walked outside, we had to bundle up and cover our faces with scarves. Our cheeks would burn from the cold and get white spots on them. The ends of our ears and nose could get exposed, so we had to be careful. The chicken barn was right near the house, so we'd shovel a path through the snow to get to it.

We were independent kids. We made our own lunches and walked together to school two miles away, five days a week. In winter, we used our toboggan that sat four, so one kid would pull it. That was our transportation to school. We liked going to River Park in winter because it had a toboggan slide, and we'd all pile on a ten- or twelve-seater and slide down the icy chute beside the Red River. We also had a sleigh with runners. No car in those days. In the winter, we'd look forward to summer and by August we were look-ing forward to winter and then by March we'd really look forward to summer again.

My younger brother Denny played hockey and was a goalie on the school team. Dave and I had skates, but we didn't get that much time to practise. I had a job, thanks to Denny—he got me started selling magazines. I was seven.

How it happened was that a guy in a truck came by the school to promote a magazine. He saw Denny, who was six years old, and he gave him a canvas bag with five copies of *The Saturday Evening Post*. The guy told him to give it a try selling them. My mom bought one copy for five cents, but when Denny realized he'd need to walk to the neighbours' to sell more he said he didn't want to do it. The guy in the truck figured Denny was too young, so I offered to try: "Leave five more copies and I'll sell them." A couple of weeks later, I said, "Leave twenty-five." On Tuesday, the guy dropped off the extra copies, and I sold them that week.

Four afternoons a week, Tuesday, Wednesday, Thursday and Friday, I'd be selling those twenty-five magazines. Sometimes I was still selling them at night, walking along the streets or travelling on the streetcar into Winnipeg. By Friday night, I'd have a hard time selling them because it was the end of the week, and a new edition would be coming out on Tuesday again, but soon I had twenty-five regular customers.

Then I got wind that the magazine was offering a contest. It was going to run for ten weeks, and whoever sold the largest number of extra copies over their set number (in my case twenty-five) would win a bicycle. The next Tuesday I took two hundred copies from the man. I gave Denny the twenty-five copies for my regular customers and told him to deliver them, which he did. I split the profits with him for selling those.

The winter cold didn't discourage me, but I knew I needed a plan if I was going to sell 175 copies. I bundled up and Tuesday filled the bag with fifty copies. I got on the streetcar and went into downtown Winnipeg. I stopped at an apartment building that had a sign on the door that read "No Soliciting." I went inside anyway. I started knocking on doors, and when the residents saw this little kid with a big canvas bag of magazines they were curious. I'd brought along the pamphlet about the bicycle to show to my potential customers, and explained that if I sold a certain number of magazines, I'd win the bike. I must have looked pretty cute to these strangers, with my desire for the bike obvious, and they probably felt sorry for me on those cold nights. Whatever the reason, that first night I sold all fifty copies. The next night I took fifty more and went to the next apartment building with its sign on the door that read "No Soliciting." Every night I took fifty copies and went out in the cold to sell them.

At the end of the contest I was declared the winner. I had won the bike. I was excited and couldn't wait to get my reward. Unfortunately, things didn't work out the way I had hoped. They shipped the bike from the States, where kids who sell magazines are older and bigger. The bike was for an adult! Big balloon tires, and impossible to ride—I had to ride it sideways with my legs through the crossbar. I quickly figured out that this was a useless prize as it stood, but that I could make it useful if I sold it for enough money to buy a proper bike, so that's what I did. I got $12 for it and bought a CCM bike for myself.

Over two years, I made a little extra money through that job. The other popular magazines at the time were *Liberty* and *Ladies' Home Journal*, and they asked me to sell them too, but they were too heavy. I only sold *The Saturday Evening Post*. I sold them for five cents a copy and I got a cent and a half commission, so I'd sell two hundred copies and make $3, which was big money. In the winter, I went into the city and sold them to the people in the apartment buildings. In the summer, I sold them on the streets, house to house.

Three years after I started selling, I won some recognition for my work—in all of Manitoba, I was the leading salesman for *The Saturday Evening Post*. I was just ten years old. I kept taking 175 copies whenever they had another contest. The second prize I won was an electric clock and then some plates.

My mom was also adding to our dishware. During a six-week promotion, she would go to the movies every Monday night and get a dish or a bowl from the usher. We were always happy to try out for contests and promotions. The Famous Players theatre offered another opportunity for a prize when I was about eight years old.

I discovered that I was a really good bolo bat player. The bat was a round wooden paddle with a string attached to it and a rubber ball on the other end. I'd bat the ball over and over, and the ball would always hit the bat. I knew I was good at it because at school, no one could compete with me, so when I heard that Famous Players was holding a bolo bat contest, I signed up.

The contest brought together seven of the top eight- to ten-year-old kids in the area. We went into the theatre and walked up onto the stage to begin batting the ball. I kept batting and never missed, but kids started dropping out when they missed. I kept my eye on the ball and hung on until it was down to one other kid and me. I figured if I could go a little longer, I'd beat him, so I just kept going. Finally, the kid missed the bat and I was the champion. I'd won a big hardwood trophy that I put in my bicycle carrier, and I rode home feeling pretty good.

I was still selling magazines during my bolo bat days, and earned enough to buy a BB gun. This purchase led to me getting in trouble. My brothers and I put a target on a fence post in the trees, so I started shooting at it. I didn't realize the pellets were going past the target, across Osborne, and into a guy's plate-glass store window. I felt terrible when Mom had to pay the man $12 for the damage, and I felt worse when the police took my gun away and sent me to the juvenile detention place in Winnipeg. I had to sit in an office and wait for my punishment. A man came in and unlocked a cupboard and brought out a strap. He hit each of my hands a few times and told me to get home and stay out of trouble. The strap stung, but I didn't cry. We were just kids. We never meant any harm to anyone.

I soon got right back to work with my brothers. We'd spend a couple of hours looking in garbage cans for empty Beehive corn syrup containers. We'd peel off the labels and send them into the Toronto Maple Leafs' office or to the Montreal Canadiens to get a signed photo of the players—we got pictures of Syl Apps, Turk Broda, Harvey Jackson, Charlie Conacher and Buzz Boll. I had quite a few.

We loved hockey. Like I said, I didn't have time to play because of my sales job, but I was able to listen to the games on the radio. Every Saturday night, my family and I would sit in the front room and we'd turn on Foster Hewitt and listen to the hockey game.

We always had plenty to do, whether playing or working or going to school. Christmas time was never a big thing. We didn't go to church or anything. We'd put up a tree and decorate it, and get presents. We always had a turkey dinner. We also believed in Santa Claus for a while.

I didn't read a lot. Didn't even read *The Saturday Evening Post*. The subjects I took in school don't stand out in my memory except for math and typing: I was the fastest typist; I could type faster than the girls. I caught on to it and was good at it.

I never thought about my future. I guess I imagined we'd always live in Fort Rouge, but World War II changed things, and when my dad got involved doing his part for the war effort, we had to move to the West Coast.

When we packed up for Vancouver in 1942, I had to leave most of my possessions behind. The bolo bat trophy was too heavy to carry, and I couldn't bring my bike. We brought the electric clock though. We only took what we could put in our duffel bags. I was still selling magazines practically up to the day we left Fort Rouge. The guy who first got Denny and me into it wanted the list of my two hundred clients, but I told him I only had twenty-five regulars; I peddled the other 175. He was shocked I had done that well.

World War II was underway, and my dad had joined the Air Force in Winnipeg as part of the grounds crew. After a while, he got shipped to Sea Island in Vancouver, so we were going to the coast to be with him. My mom had to bring the six of us kids on the train by herself. We were excited to be on our first train ride. It took three days and a couple of nights to get to Vancouver. It was a long and noisy ride on that steam-driven train. My youngest brother Herb was around seven years old at the time, and I think he slept most of the way. The rest of us slept at night as the train rumbled along, and during the day we'd walk up and down the carriages or play cards or look out the window. My mom had brought food for us, so we ate sandwiches most of the time. I thought about the big wooden trophy that I won at the bolo bat competition and about my CCM bike and how I had to leave all that behind. That's pretty much all I remember of that ride to our new home in Vancouver.

When we came through the Jasper area in Alberta, we saw the fenced Japanese internment camps and their beautiful gardens. The men and women were leaning on the fence, watching the train pass by. We had no idea why they were contained or what that meant. Later we learned that the Canadian government figured the twenty-two thousand Japanese Canadians were enemies of the state after Japan bombed Pearl Harbor in 1941. The authorities took all the Japanese citizens' possessions and scattered the people around the country to these camps. The internees along the BC–Alberta border tried to make do: they built tidy houses and gardens, and lived this tough life until, at the end of the war, many of them were told to go east or go back to Japan.

The other important point about Jasper is that my future wife and her family were living on their farm nearby in Lucerne. I had no idea this trip to the West Coast was leading me right to her, but before we met I had more work to do.

MY MILL DEBUT

I N THE SPRING of 1942, when my family moved from the farm in Fort Rouge to Vancouver, we didn't know what to expect. Things started to get better for the family with my dad's regular paycheque from working with the land crew in the Air Force. He was on leave when we arrived in Vancouver, so he was able to meet us at the train station on Main Street.

When we got off the train, our dad was there to meet us. There we were, Mom and the six of us carrying all of our stuff in paper sacks and duffle bags; we didn't have proper suitcases. As I walked up to Dad, I was thinking that our new life in Vancouver might make for a great adventure. He said our new home was just up the street at 5th and Main. We crossed Main Street and walked up to our rented house at 128 5th Avenue. A Japanese family owned the house, and there was still a young fellow living in the attic when we got there. I guess he got shipped out to the internment camp soon after we arrived because I don't remember seeing him again.

The house was small, so there wasn't much to explore indoors. We put all of our stuff away in the rooms and wondered what next. Since it happened to be a Saturday afternoon, Dave, Denny and I asked Mom if we could go see a show.

Mom in Vancouver, 1942. She always wore an apron at home.

She probably gave us about thirty cents for the three of us to see the show. Movie tickets in Winnipeg cost ten cents, and we figured it would be the same in Vancouver. Some theatres would give you a free ticket for the matinee if you brought a spoon or fork or anything metal to add to the war effort. Armed with our dimes, my brothers and I walked up to Broadway and then carried on west along to Granville and to the Stanley Theatre.

On the way we passed a few stores and looked at the displays. Eventually, we walked by a bowling alley. The bowling alley wasn't open yet, but we stopped and looked at a card in the window: "Pin Setters Wanted." There was an old guy out front sweeping the sidewalk, and we asked if he thought we could do the job. He looked us up and down and said, "Do you know anything about setting pins?" Dave and I glanced at each other and said, "Sure we do!" Denny just stood there, not saying a word, probably wondering when we had learned to set pins in Fort Rouge. We had never set bowling pins in our lives, but we knew we could figure it out. He said to come back later when they opened. We asked him for directions to the movie theatre and said we would see him soon.

We entered the theatre and took our seats with all the other youngsters there to see the Saturday afternoon picture show. The lights dimmed, the screen lit up and we felt that familiar feeling of excitement to see a world different from our everyday life. That first picture show we saw in Vancouver was *Sergeant York*, starring Gary Cooper. In one scene I recall him shooting at turkeys when they'd raise their heads. *Gobble gobble gobble.* I can still see him shooting those turkeys to this day!

We stopped by the bowling alley on our way home and met the boss. He asked, "Do you boys want to start working tonight?"

We said, "You bet we do!"

"We only need two of you and that little guy's too young."

We told Denny to go home and tell Mom where we were, but he just stood on the sidewalk and cried, "I don't know where we live!" You have to remember we'd just moved into our new house that day. We showed him the general direction and then sent him on his way. By that time, Mom had already started looking for us and eventually she found Denny, and he told her what we were doing. We were working!

I still remember the name of that bowling alley—Chapman's Lanes. There were sixteen lanes on two floors, all five-pin bowling. I got lanes 9 and 10 on the bottom floor in the middle of the place. Those were the good alleys because people always started on them. So there I was, sitting between the lanes as the balls would fly toward me. I could pick three balls at once and put them on the rack and they'd roll back to the customer. I'd set three pins at the same time and then plunk down the other two. It was pretty noisy back there as the balls came whizzing down the lanes and I had to move fast, but I soon got used to it.

That first night we finished work at midnight and walked back to our new home, happy that we had found work on our very first day in Vancouver. Dave and I worked at the bowling alley for about a year. It wasn't really hard work. You had to set the five pins in a V, jump down and put the balls in the rack to go back down the alley. Luckily they were small balls not the big ones with holes for fingers. Those weighed much more.

We got paid once a week. The bowling alley would check the sheets and tally up how many strings were played. Strings are games. Each game had ten frames. We earned four or five cents a regular game. Players would send a tip of a quarter down the alley, so we always made a little more than the basic rate. During league play you could make a lot more, up to $2.40 a night setting pins. That was good money back then.

On Saturday mornings at 10:00, we'd set pins for the blind when they had the lanes all to themselves. I remember watching them feel their way along, holding onto a railing near the lanes as they threw the ball.

I took that job even though I had never set pins before in my life, but I just knew I could learn. I have to admit that if someone ever said to me, "You can't do this, you can't do that," I'd never take no for an answer. Someone might tell us we shouldn't do something, but I'd say, "Well, let's give it a try." Over the years, maybe we could have held back sometimes, but for most of my life it was full steam ahead.

We lived at the 5th Avenue house for only a short time, not even a year. Dad eventually found our second home farther west in Kitsilano at 1785 West 2nd by the Seaforth Armoury and the Burrard Bridge. The landlord of that house also had a big shed out back. This is where we found our next job—baling shavings. The fellow would come around with his pickup truck filled with wood shavings. He had a contract at a local sawmill where they made shavings from the side slabs off logs. He'd go there with his truck, get a bunch of shavings and dump them into the big shed in the backyard.

He showed us how to make bales. We'd put a wooden board on a rack, then put shavings on the bottom, a board on the top and two boards on both sides. Pull a couple of levers and the wire machine would automatically wrap wire around it—two wires per bale. These shavings would be sold to farms, especially chicken farms. We couldn't make enough of the stuff because there was such a big demand for wood shavings. We'd run the baler and stack the bales on the truck and earn a few more cents. We went to the mill and watched how they made the shavings from the slabs. Some of the slabs were cut up for firewood. Other slabs were used to make chips and we'd have piles of them. You couldn't make any wood from those chips; they went to making paper. Shavings went to chicken and turkey farms. Nothing much was wasted. There were no chippers in those days.

In the early 1940s, part of Vancouver was like an isolated lumber town with a lot of sawmills. I remember them being on every block at False Creek and over in North Vancouver. The air was thick with smoke from burning wood and coal. The mills had beehive burners to burn the waste.

During the 1930s, many woodstoves were converted to sawdust, using hoppers to feed them. That was a big funnel on the side of the stove filled with sawdust that fed the flame below. There were also sawdust burners to feed a furnace. To feed our furnace and woodstove, we'd partner with a neighbour to get firewood. We used a saw, bucking wood to cut the timber that we'd collect from a small forest near our house. I can still remember the slapping of the belt on the circular saw as we cut the wood into stove lengths. After all of my years working in mills with wood, I'm happy to say I still have all of my ten fingers. Even today, the men who make wood shingles have a greater chance of losing a finger since the shingle is cut so close to the moving saw blade.

When I wasn't working, I was in Grade 9 at Vancouver Technical School near Hastings. Mostly what I remember about that was it was a long way away. By then I had another CCM bike and would ride back and forth to school from Kits.

My dad wasn't comfortable living in the city. He wanted to be back on the land, so not too long after we arrived on the coast, we moved several miles out of Vancouver to Haney, which is now the downtown core of Maple Ridge. We rented a large thirteen-room house at 216 and River Road, but back then it was 5th Avenue and River. If someone today calls it 5th, you know they've lived there a long time. It felt great to have all that space after living in a four-room house. Each of us had our own bedroom.

We lived about a year in the big house and then moved to 14th Avenue South. My dad was now able to rent a house with a piece of land, so we had a few acres with chickens. We soon packed up and moved again, this time to 14th Avenue North. After all those rental houses, we were finally able to buy our first home. The house was pink stucco and located on a hill overlooking ten acres. We got the house through the Veterans' Land Act. The government loaned ex-servicemen the down payment and helped with the cost of livestock and equipment. Although Dad had never left Canada during the war, he was still a veteran, so he decided to make the most of the opportunity.

Out back of the main house was a small three-room bunkhouse. Dave, Denny and I slept there. We had an outhouse back there too. The farm had two acres set aside for growing hops to make beer. Dad didn't drink, so he had us help him clear out the whole area.

Our house in Haney on 14th Avenue North.

Most small farmers had to have a second job to supplement the farm income. Dad had his job at the Maple Ridge Lumber sawmill, which helped add to the money we earned from our chickens and half a dozen cows. One of our chores was to milk the cows. We sent our milk to Fraser Valley Milk Producers' Association. We'd put out the big twenty-gallon cans on the road and a truck would pick them up. I remember going with Dad to the local farm auctions. We'd go down with a cow and bring back some chickens or else we went with some chickens and brought back a cow.

While the kids were at school, Mom went to work in the kitchen at the Haney Café washing dishes. She wasn't ashamed to work in a diner kitchen—it was money. If someone came in to the restaurant and was deaf, she would communicate with them and take their order using sign language. I have no idea how she learned, but she knew sign language.

I went to Grade 10 in Maple Ridge School until I was fifteen and in November left to find work. My friend Billie Bird had worked at the Hammond Cedar Mill, but it had shut down for repairs, so he was now working at the Youbou mill on Vancouver Island. The man who owned the Hammond mill also owned the Youbou mill. Billie thought I could get a job there. I had nothing to lose; I didn't know for certain I could get a job, but it was worth a try. I borrowed some money from Mom and took the bus to the ferry that would take me to Vancouver Island. Six fellows on the bus were also going to Youbou. They'd already been hired. I hoped I'd get lucky and this would be a new work adventure.

Youbou was a mill town on Vancouver Island located eight miles west of Lake Cowichan. It got its name from two employees of the Empire Lumber Company who operated the first sawmill there. Mr. Yount was general manager and Mr. Bouton was the president. Youbou was famous for having the last handset bowling alley in all of North America. If I couldn't get a job working at the mill, I guess I was prepared to work as a pin setter if I had to.

When I first arrived at Youbou, I wandered in with the six fellows, but the foreman said they didn't have any work for me. I panicked for a bit because I didn't have any money. I decided to stay and wait for a chance to get work. Every day I would line up with the other men to eat at the bunkhouse kitchen. I'd ask at the office, "You got any jobs?" They'd say nope. I'd been eating meals for four days when a man came up to me and said, "Do you work here?"

I had to say no.

"Well, how are you going to pay for all this food you're eating?"

I said, "No job, no money. Give me a job and I can pay it off."

I was so small they didn't think I could do much of anything at the mill. But the next day the man called me in at lunchtime: "You're going to work after lunch so you better eat good."

I know they made up a job for me so I'd pay my bills. I had to go where the two certified Pacific Lumber Inspection Bureau (PLIB) men were grading the lumber. I had to straighten the board as it came down the belt past the two graders. They could have done it themselves, but they'd made this job for me. When the belt stopped and no boards came to be graded, I saw the men down the line tying smaller boards, like one-by-fours, for car loading. I watched and then asked if I could do that, and they said sure, so I learned a new job. I also learned to drive a forklift and a lumber carrier.

After a few days, at the end of January 1944, the foreman gave me my pay envelope. The two graders teased me and said, "That's your cheque and your pink slip. Open it." I was shy about opening it there. I waited until I was in my bunkhouse to see if I'd been let go. I knew it might not be enough to pay my bill and get back home. My four days of room and board had been deducted, so it wasn't much of a cheque, but I hadn't been fired—I was still an employee.

During my time at the Youbou mill I was paid fifty-two cents an hour—$4.16 a day. Out of that they took $1.50 for room and board. The food must have been pretty good because I grew three or four inches and put on twenty pounds. I lived in a bunkhouse and slept on a wooden cot. Showers were down the hall.

I met a French guy about my age and size, so we became friends. We used to work forty-eight hours a week, but Saturdays we worked only until lunch. Saturday afternoon and Sunday we got off. One Saturday, we saw a steam freight train with the boxcars loaded with lumber from Youbou and heading for Victoria, and decided to have an adventure. We jumped on the side and hung onto the ladder. The rain was pelting down, so we got soaked. When the train stopped to get water for the steam engine, a trainman saw us. We thought he was going to tell us to get off and walk the whole way back to the mill, but he was a young guy, so he helped us. He told us to go into the engine room where the fire box could dry us out. We almost climbed into that fire. Sure felt wonderful to be warm.

When we got near the yard in Victoria, the engineer slowed down so we could jump off. They couldn't take us right in because it was against the rules. We had got dry completely, but it was dark and still rainy. We had no idea which way to go, so they gave us directions to the main part of the city. We were wandering on one of the main streets when the police stopped us and said there was a curfew. We explained we were from Youbou, showed them our paycheques that we hadn't cashed yet and said we had no place to stay. We were only fifteen, and we had to be sixteen to get a registration card to identify us, so all we had for ID was our paycheques. Without ID we'd be vagrants and they'd throw us in jail. The police phoned around and found us a boarding house and we paid $3 for the night and then hopped a bus back to the mill.

I stayed at Youbou until May 1944. When I got back home, my mother didn't recognize me because I'd put on some weight and grown taller. I looked different, and my life was changing too. My time at Youbou was short, but the work in the mill set the course and direction of my life.

I got work right away in the planer mill, tying up lumber at the Hammond Cedar Mill, which later sold to B.C. Forest Products. My brother Dave was a mechanic's helper, and Denny worked in the boiler room. We all worked day shift together. We carpooled with a guy in a truck who'd pick us up at the corner of 14th Avenue and Dewdney Trunk Road. We paid him for the daily ride. We got home around 4:30 in the afternoon.

Forty women and men who couldn't make it in the army were working with me in Hammond in the planer mill. The women were grading and tying the lumber like at Youbou. They had their own lunchroom and washroom. After the war, the company didn't replace the women when they left, but they'd only hire men because

I bought my first motorcycle in 1944, a pre-war Harley army bike.

there were more guys looking for jobs, so gradually they had fewer women. I stayed at Hammond for four years.

I got my first motorcycle, a reconditioned pre-war Harley army bike, in the summer of 1944 when I was fifteen. The motorcycle shop was in Vancouver on Broadway. I didn't know how to ride it, so a friend of mine who had a bike rode the bike home with me on it— an awkward ride because he was way up front on the tank. I had to learn how to drive it. I would say to myself, "Shift gears, heel, toe and away we go."

To get my licence, I had to drive to Mission, eighteen miles away. I had to take my dad to sign for me. The bike only had room for one person, so I was sitting on the tank and Dad was on the seat. Not a great way to travel. I wrote a small test and then the instructor told me to go up a hill and turn around and then come past him at the bottom of the hill. That was the test. I passed.

Dad wasn't fussy about the bike. He'd often get a ride from a guy in a truck to and from Haney Mill and then get dropped off on the corner. One time, I drove up and asked him to hop on the back. He

My parents in their later years.

hadn't sat down properly because it was cramped on that one seat, and I started out too fast so he slid off and landed on the ground. Another time, I gave him a ride to the Haney Mill on my way to Hammond. The road was icy, so when I got to the corner to drop him off, the bike slipped out and landed on the side. Dad fell off again and this time got his pants wet, so he said he wasn't going to go on the bike ever again.

Our parents lived at 14th Avenue North for over twenty years. Dad died in Langley at the age of sixty-nine in 1969. After Dad died, our brother Dave bought the house from Mom, who then went to live in a seniors' home. Mom lived to the age of seventy-six and passed away in 1981 in Maple Ridge.

As I said, I've been lucky never to lose a finger working in the mills, but I still have a couple of scars from an accident with my motorcycle and a friend's car.

The first accident involved that first army bike. I broke the drive chain, so a friend with a car said he would take a cow chain and attach it to the bike to pull me home. We got the chain on the bike and I was sitting on it, steering. My friend was pulling me along when we hit some mud and the bike fell over. There he was, dragging me in the mud and along the gravel road for a ways before he realized what had happened.

When I looked at my knee, all I could see was a huge bloody gash filled with gravel. My friend took me home, and what I'd call a real horse doctor came over and was grumbling while putting metal clamps on my knee to pull the skin together. He'd been drinking and he wanted to go back home and finish his drink. The clamps were useless. In the morning they had fallen out because my leg had swollen up so bad the skin had stretched and pulled the clamps out. That doctor had barely cleaned out the wound, so my mom put hot cloths on it to try to remove the stones. We did that with the cloth for two weeks. I was off work the whole time. When I finally got back to work, the doctor, who had an office across the street from the mill, kept putting bandages around it to keep the skin closed. He said I wouldn't be able to bend it again. I hobbled around on a stiff leg for a couple of months.

By month three, I looked out in the yard and there was my motorcycle leaning on its stand. The chain was fixed, and it looked like it was waiting to be used. I wanted to take it for a ride, but I didn't know how that would work with my stiff knee. I stood beside

Willie Lykowsky and Vic Rempel on the left. Phil Stolarski behind me on our new Harleys in 1947.

the bike and thought what the heck and jumped on. I normally had to bend my knee to work the clutch, but this time I sat way back on the seat and I kicked it down with my heel. I tried working it with my stiff leg and that seemed to be okay. I was so excited to be back on my bike that I guess I stopped concentrating on my leg because, by the time I got to my friend's house in Haney six miles away, I'd been using the clutch so much I didn't realize I had started bending my knee again. Maybe it was mind over matter.

The next bike I bought was a brand new 1946 Indian Chief. No Harley-Davidsons had been made during the war, so I bought the Indian Chief. This was a long saddle seat so I could sit way back, but I could also get another person on it if I sat forward. A nice bike. Willie Lykowsky was a good friend at the mill, and he'd ride around on the back of my bike. He wanted to buy one of his own, so in 1947 we both got new Harleys.

Willie had to walk on the ball of his foot because the tendon in his right heel was too short. This made it hard for him to kick-start

I'm dressed up for a ride on my 74 Overhead Harley from 1947.

his bike. This wasn't a problem when we were on our own, but if people were around he'd get embarrassed about struggling to start his bike. One time outside a restaurant a few guys were standing around looking at the bikes, and I could see Willie was having trouble turning it over, so I said, "I'll kick it over for you." I stretched out my leg and did it from the side of his bike. Easy fix.

My next bike was a 74 Overhead Harley-Davidson from 1947.

That year, Willie and I joined the Greater Vancouver Motorcycle Club that Trevor Deeley was running as head of the twenty-member group. The club rented a building on Broadway for our bi-monthly Wednesday night meetings, where we planned events and rallies. Throughout the summer, we would take weekend trips and travel a few hundred miles out of town, up to the interior of BC and to Kamloops or Prince George. Married guys would bring their wives, and it was a fun way to see parts of the province. We also had smaller day rallies of 100 to 150 miles, where we'd head out of town and stop at a restaurant or a picnic ground for lunch.

One of the events we planned in Burnaby was a Saturday evening scavenger hunt. Willie and I got the list of a dozen items that we had to collect before the 9:30 deadline—we had two hours for the hunt. We read through the list and figured most of the items wouldn't be too hard to find, except for the last one: a pair of ladies' panties. Willie hopped on my bike closest to the headlight and I rode behind. We drove over to a local gas station and sifted through the garbage for a used No. 4 spark plug, took our pencil and checked that off the list. We drove down Kingsway, not sure where we were heading, when we noticed three teenage girls standing outside a burger joint. We pulled over and they approached the bike.

I explained that we were on a scavenger hunt and needed to get a pair of panties. After they laughed, one of them asked to see the list. She read through it, saying, "I've got that. And that. And that." She lived about two blocks away and would be happy to get a few of the items from her house. Willie got off the bike and stayed with her two friends while I drove her back to her house. A few minutes later, she came out with a brown paper bag. "I've got a spool of green thread, a safety pin, a darning needle, a button, a roller skate key and the panties."

She hopped back on the bike and we returned to Willie and her friends. I knew where she lived so I told her I'd bring everything back later that night. We thanked her and headed to a friend's house, where we got a deck of cards and a matchstick. We stopped at a corner store for a chocolate bar, which we ate, and crumpled up the wrapper and added it to our haul. The photograph came from my wallet and Willie took off his sock. We now had all twelve things, so we raced back to the clubhouse to see if we were first.

We handed the bag to Trevor Deeley.

"How'd you do?"

"We got everything."

He took a careful look inside the brown bag and said, "Looks good."

We waited to see who else had been lucky. Turned out, the item that caused the biggest problem was the ladies' panties. A couple of guys had tried to take a pair off a clothesline, but they couldn't reach that high and gave up before someone called the police. Most of them only got half the stuff, but when the sixty-year-old Australian bachelor came in, he proudly shouted, "I got panties!"

He brought out the pair he had found (he never said where) to make his list almost complete. No one could tell what colour they were because they had so many holes—they were completely moth-eaten. He didn't care: he had come in second. No one got a prize, but we all had a big laugh over the "holy" pants. I kept my promise and returned the brown paper bag to the girl. No one except Trevor had looked inside, so I couldn't say what shape the panties were in or what colour, but I can guarantee they didn't look like the pair the Australian found.

I had fun on the rallies and attending the club's meetings, but I was a little unlucky on that Harley. One night, coming home from work, I hit a bump on the road and flew over the handlebars. I landed on the left side of my face and scraped the backs of both my hands. My left cheek was open like a flap, my forehead had a gash in it above my eye and my lip was cut. I don't remember how I got to the doctor's, but when things settled down I had twenty-one stitches on my face and both hands were bandaged. We never wore helmets in those days. I usually wore a white cap, not worth a dime for protection. I was glad I hadn't broken any bones or lost the sight in my left eye.

I was eighteen when I *did* lose something precious on my Harley. The legal drinking age was twenty-one, but the guys in the mill always invited us younger workers to join them for a beer. On this particular night, we were going over to someone's house, and they nominated me to go to the beer parlour on my motorcycle to pick up two cases. I waited out front of the beer parlour until I saw a mill-worker I knew and asked him if he'd do a run inside for us. He agreed, so I gave him $5 for the twenty-four beers. A few minutes later he came out with the two cases. I balanced them on my bike's gas tank and headed down the main highway, which was just a two-lane road.

In a small town like Haney, the police knew most of us and would cut us a little slack about drinking. If we were at a park at night, they'd tell us to go home, but they never took away our beer. I was heading down the highway with the beer on my tank when I noticed a police car driving the opposite way. I tried to look casual and older, but the police car turned around and came up behind me. I hated to pull over but I did. I recognized the policeman; I knew him well, but he didn't look friendly, more like sad. He was shaking his head when he said, "What're you doing driving down the highway with that beer on your lap?"

"I'm taking it to my friend's."

"That's not what I mean." He glanced back at the police car, and I noticed another person in the passenger seat. "I gotta ask to see your driver's licence and registration."

I had to put the beer down on the ground when I showed him my papers. He handed my licence and registration back to me but pushed the beer away with his boot.

"See that guy in my car? That's the police commissioner. I'm really sorry, but I have to confiscate your beer." As he leaned down to pick up the two cases, he said quietly, "What're you doing on the main highway? Why didn't you drive down the back road?"

I had to pay a fine and watch him take away our Saturday night beer. The traffic stop had cost us a day's pay and spoiled our party but at least I wasn't injured. I had suffered a few scrapes on that bike, but the bigger accident came when I was a passenger in a car.

On another Saturday night a friend, Alan Anderson, was driving five of us back from a dance at Pitt Meadows. We always raced from the dance to the Haney Café to get one of the few booths in the restaurant. We were speeding along the Lougheed Highway when the car blew a tire. The car flipped, and I flew right out the canvas roof. No seat belts at that time. I remember it hurt when I landed on a rock in a driveway.

A guy coming along the highway from the dance took us to a local doctor, where I was patched up and sent home. When I got into bed I had a hard time sleeping because my back hurt so much. Suddenly, in the middle of the night, my brother Dave woke up and yelled, "Chick, are you asleep? I had a dream that we were in Alan Anderson's car and we had an accident."

I guess Dave had been knocked on the head during the accident and had lost some of his memory. I said, "That was no dream." To prove my point, I said, "Come and take a look at my back."

Dave got up and I rolled over and showed him my back. By that time it had swollen to the size of half a football. He said, "I'm going to get Mom!"

When Mom came from the main house and took a look at my back, she told Dave to call the doctor. The doctor, an old army doctor, came to the house, took one look at me and said, "Broken back. Get him to the hospital."

And so there I was at Royal Columbian Hospital in a ward with four beds. Funny thing—I knew the guys in the other beds. One of them was Bob Poulevard, a sawdust truck driver from Surrey whose truck got stalled on the train track and a train hit him. He lost his leg. The other guy was from Fraser Mills driving a lumber carrier around the yard when it flipped and wrecked his shoulder. The third guy had a broken leg after his motorcycle fell on him.

The doctor was standing over me and said, "It's a hematoma and the blood has to be removed." He added, "I hope you don't mind, but we have some student nurses who will be watching the operation." Sure enough, a row of young nurses stood around the bed.

At first, the doctor tried to remove the blood with a large syringe, but the blood had already congealed, so he took a scalpel and carefully sliced into the hematoma. I didn't feel a thing. The blood started gushing out, and the clots were coming out in chunks. One by one those young nurses left the room.

These days, the only residue of discomfort from those early accidents is that the knee and the back get a little achy, but all in all I've been lucky.

The doctor had been telling me for months that I needed to get rid of my motorcycle if I wanted to survive my youth. He also warned me that, after wrecking my back, if I took another fall I could hurt my back for good. I didn't sell it right away. In fact, my motorcycle soon proved invaluable.

THE GIRL OF
MY DREAMS

I N JUNE OF 1948, Hammond Mill shut down when the mill floor was flooded. The Fraser River had been rising since the Victoria Day weekend, but no one was really worried until a couple of dykes broke. By early June we were in trouble. The water came up about twenty feet over the banks of the river and broke through several dyke systems. We had water covering over a third of the Fraser Valley floodplain from Chilliwack to Mission. The heavy snowmelt came late, and we got warm weather that week; the snow melted all at once, forcing thousands of people to get out as fast as they could. Several thousand homes got destroyed in that flood.

My brothers and I volunteered to fill sandbags over in Pitt Meadows. The phone lines were out and the major roads were flooded, so communication between communities was cut off. Some of the smaller roads were still accessible by motorcycle, so the men in the Municipal Hall asked if I would dispatch for them. I was still riding my 1947 74 Overhead Harley, the big one, so I agreed.

The municipal men were getting bulletins from helicopters overhead about how the flood was doing, so my job was to take the bulletins to different dispatch stations. The helicopters were following the progress of the flood, and they needed to pass information

An overhead view of Gossip Island, BC.

to the flood watchers—mostly local farmers keeping an eye on their property or men at gas stations on high ground throughout the district. I'd get the report and jump on my bike and ride by Whonnock toward Mission to let them know the situation. For a week, I kept busy three or four times a day riding along those back roads. Our home was on the top part of 14th Avenue, so we were okay, but the flood still affected the way we got home. The river took a month to lower down to normal.

A couple of years before the flood, in 1946, the Hartnell family sold their Hammond sawmill to B.C. Forest Products. The Hartnells wanted to stay in the mill business, and they liked the Gulf Islands, so they decided to build a mill on forty acres over on Gossip Island. My friend Cam Ritchie had worked at Hammond, so when they started building the mill, Doan Hartnell hired him. By 1948, the mill was built and running. The flood was still on, and I wanted to get back to work to get a change from dispatching bulletins, so

I phoned Cam and asked him if they needed men over there. Cam checked for me and phoned back. "Come on over. I got a job here for you."

Gossip is a small southern Gulf Island beside Galiano Island in Georgia Strait just outside the entrance to Active Pass, where the ferries run between the Lower Mainland and Swartz Bay on Vancouver Island.

On the ferry on my way over to Gossip, I saw the effects of the Fraser River flood. Cows and other dead animals were floating through Active Pass near Galiano Island on their way to the open Pacific. Sheds and debris were mixed in with driftwood and were still floating by when I was at Gossip.

When I worked at Hammond, I'd been tying lumber, but I got a new job at the Gossip mill. I was to oil and grease all the machinery and light the burner in the morning. I started work early and ended around 3:30 p.m. on an eight-and-a-half-hour shift. First thing, around 7:00 a.m., I'd get the fire going by pouring a pail of diesel oil into the beehive burner through the bottom door. The fire would still be smouldering from the night before. I would oil the edgers and the head rig where the log gets sawed on the carriage that goes back and forth, and I'd oil the chains and the bearings, greasing up places I could only get into while the machines weren't running. If anything broke down, I'd help the millwright. He was the guy who kept the machines working; he knew how to fix them. He had more advanced training than an oiler.

The mill was pretty noisy, so whistles were the best way to communicate. Four whistle blows meant there was a breakdown and brought the millwright over. Three whistle blows would be a call to the electrician and two was for a foreman. One whistle at 9:00 a.m. was our ten-minute morning break. Then another whistle at noon. We usually got half an hour for lunch. Some of the guys in the yard didn't wear protective gear for their ears, but if you were around the planers and head saws, you had to. Most good workers used earplugs or earmuffs. Your ears were protected but you could still hear the whistles.

Sleeping arrangements at Gossip were similar to Youbou—I slept in a bunkhouse with about ten other guys. Some workmen came over from Galiano by boat to work at the mill. About forty workers ran the whole operation. At Hammond, I was paid seventy-two cents an hour, but at Gossip I earned ninety cents an hour, plus I got some good experience.

The system in the mill was pretty standard for that time. Loggers brought logs in from the camps on Vancouver Island. The logs came in on booms tied up with boom chains and were towed in the water over to Gossip. We didn't cut any trees on Gossip. Lots of activity happened in the water: the logs came in on booms and the lumber went out on barges.

Back then, the log waste came into the beehive burner, but now the waste goes into chippers to make paper. The logs came into the mill on a jack ladder and were raised to the top of the mill and onto the log deck. They roll over the carriage that saws them up. We had circular saws then, not band saws like today. The lumber goes down into the mill and the edger takes the sides off them. Then they go out onto the green chain to be pulled off on certain loads and to be graded. Gossip was working on fir and hemlock, so we'd make different sizes in the softwood: two-by-fours, two-by-sixes, two-by-tens, and different grades. Some mills might classify No. 1 as construction grade, No. 2 as standard, No. 3 as utility, and No. 4 as economy. Or you'd call it high- or low-grade lumber.

For example, Hammond was a cedar siding mill, so their red cedar was high-grade lumber used for its look. Bevelled siding was their main product, bungalow siding for decoration or clear lumber for siding. Hammond also produced some shiplap, the lowest grade wood used for barns or sheds. Lower grade is used in construction, like two-by-fours for framing. Youbou, like Gossip, used hemlock and fir for construction. The wood might be graded No. 1 for construction or A or B to show high or low quality. Low grade would be No. 4 with lots of knots in it. Really low grade goes into the chipper. Mills have different systems for grading their wood.

I turned twenty when I was at Gossip and made a couple of friends. I bought a .22 rifle off one of the guys at the mill, but he didn't have any shells. One of my friends had found a .22 shell, so he said, "Let's go shoot a duck." The ducks were swimming on the ocean not far from our bunkhouse. We lined up the rifle on the table under the open window and loaded the gun. I was about to pull the trigger when my friend said, "Hey, I'm pulling the trigger cause it's my bullet." I said, "I'm doing it because it's *my gun*." He shrugged his shoulders and said, "Okay."

I saw a couple of ducks floating not far from us, so I took aim. When I pulled the trigger, there was a flash and the shell casing exploded! What the heck had happened? I put my hand up to my face because I saw blood. I cried out, "I've been shot!" The chef had heard the gun go off, and he heard me. He came running and handed me a towel to stop the bleeding. When I took the towel away from my face, the two guys stared at me and said they couldn't see anything. The chef took another look and said, "There's a little scratch on your nose. Just a graze. You'll be fine." When we realized I would live, my friend and I saw something floating near the shore, so we went out and retrieved the dead duck. I'd managed to hit the target. We figured the misfire happened either because the gun was old or the bullet was faulty.

A few years later, a big pimple came out on my nose and wouldn't go away. I went to the doctor and he said, "Have you ever been in a war?"

"No," I said.

"There's a piece of shrapnel in your nose." He cut it open and took out a bit of metal from the shell casing. All that time, the piece of bullet casing had been stuck in my nose and I never knew it.

I had another eventful adventure on Gossip when the sawyer invited me to his place for dinner. His family was living on the island with him, so it sounded like a nice way to spend an evening. I arrived during daylight but left when it was pitch black. I had forgotten to bring a flashlight, so I had to walk back to the bunkhouse in the dark.

The problem was I couldn't see anything. I started walking on what I thought was a trail and stumbled along until I heard the ocean rushing. I knew there were steep banks by the water, so I quickly retraced my steps—I didn't want to fall into the sea. I didn't think it was possible to get lost since the island was just a quarter mile wide and half a mile long. I knew a main road of sorts went through the middle of the island for the gravel truck, but I couldn't find the road. We didn't have any wildlife like bears or moose, only deer, so I knew I wouldn't be eaten alive, but I was still scared. I have no idea how I finally found the bunkhouse, but there must have been a light on inside. Boy, was I happy to get back. I'd been out wandering in the dark for over two hours!

Work in the mill was interesting and the money was good, especially for guys like me who didn't have the education. It's bull work, heavy work, but you still have to have some brains to do the job properly. It's a good trade. I got along with everybody but didn't socialize too much. Most of the guys liked to head down to the beer parlour after their shift, stay out late and get up early for work, tired and a little hungover, but that wasn't my style. I didn't like to spend my money foolishly, and I wasn't comfortable going out every night, so when I got the opportunity to work at Gossip I was happy to go there to get away from the social aspect of the job. I wanted to stay in the mills and would have stayed longer at Gossip, but the mill shut down.

I'd been there a year and stayed on at the end to help clean up with a few other guys. We still had loads of lumber in different places that needed to be sold and put on the barges. Some needed planing and some was air drying. The planer kept running until the lumber was out of there, then they took out the machines. I stayed until all the lumber was cleaned up in the yard.

The Hartnell family still owns the forty acres where the mill used to be, but a company bought the rest of the land on Gossip and subdivided it into lots. A few years later, I bought two of the best lots overlooking the point, and when I owned my own mill, I built a yellow cedar log house at the mill, put it on a barge and sent it over to Gossip as a vacation cabin for my family and me to enjoy.

Years later, I was at the cabin on Gossip and had hired a guy to help me do some work on the place. We got talking and the subject came round to the Youbou mill over on Vancouver Island. He said his dad used to work there. We did the calculation and worked out I'd been there the same time as his dad. He gave me a few more details and I realized his dad was my French friend. I told him about the two of us taking the train into Victoria for a little adventure. We were both surprised to make that connection. I asked how his dad was doing and was sorry to hear he'd been in poor health for some time.

By the time I left Gossip I knew I wanted to stay in the mill business. Fortunately, another opportunity came up when Cam Ritchie called me about a job. Cam had left Gossip before me to work in New Westminster for the brothers Bob and Ian McDonald, who owned

Years later, the employees' bunkhouse is still on Gossip Island. Over my right shoulder, you can see the window from where I shot a duck.

McDonald's B.C. Manufacturing Company. Theirs was a big red cedar mill, and the brothers wanted to get into the siding mill business like Hammond had done, so they built a planing and siding mill attached to the main mill and put in dry kilns. They originally had a box factory. Cam had heard about the new mill and talked to the brothers. They hired him right away because he had done the same kind of work at Hammond. Cam called me and said, "I can get you a job at the sawmill here. Come on over and help me."

Cam and I worked well together getting the new mill going. We were making siding. I was bundling cedar siding like I'd done at Hammond. Cam had run the planer at Hammond, so he knew what he was doing in that department, and I started teaching the new guys how to tie up, package and trim the lumber. We were doing so well in the day shift that the McDonald brothers asked me to start a night shift, and so I became the night foreman to a twenty-man crew. We did the same work as the day shift. I brought my two

TOP Mom, Dad and the five boys at our home on 14th Avenue North in Haney in the late 1940s. Dave is on the far right. I'm between Dave and Dad.

BOTTOM The Stewart brothers in order of age from right to left: Dave, Chick, Denny, Sam and Herb.

brothers on to work with me on the night shift: Dave fed the planer and Sam tied up lumber. We drove in from Haney to New West together and back home together around 2:00 a.m. As night foreman, I got $1.35 an hour! A big pay raise.

My life was pretty full with work, but sometimes, for entertainment, I'd get on my Harley and go for a ride. My friend Phil Stolarski had a Harley and he worked the night shift with me. We'd get off at 2:00 a.m. and sometimes go for a short ride. After our shift, on one particular early Friday morning, Phil and I bet the guys we could get on one bike and make it to the California border by midnight. They figured we were nuts, but went along with the bet for the fun of it. Phil said, "We hate to take your hard-earned money." All in good fun.

We started our trip at the local café where we got a bite to eat. By 2:30 we were on the highway. My Harley was a two-seater, so Phil had no problem sitting on the back. We took turns driving so one guy didn't get worn out. Every hour we'd switch. I don't remember much of the sights, but once in a while I'd feel something on my shoulder and I'd look around and there was Phil's head leaning on it. I'd give him a good nudge to wake him up. Last thing we wanted was for him to fall off and slow us down.

When the California border came into view, we were sure glad to see it. In those days, the United States had border security between the states. Phil looked at his watch and yelled over my shoulder, "It's ten to midnight." We drove up to the security guy and told him we needed proof we'd arrived there before midnight. He was already handing out the next day's tickets, but he made a special one for us to show the guys back at the mill. We got off the bike and stretched, holding the ticket and grinning. We'd done it. The excitement died pretty quick when we realized we had to turn around and do it all again.

We rode back to Oregon and got a hotel room for the night. The return trip took two days. We got back Sunday evening, which worked out fine because our next shift was Monday around 4:30 or 5:00 p.m. We got in to work and showed the guys the ticket. They couldn't believe it. We'd won the bet. They were happy to pay up; we deserved it, they said.

That twenty-one-hour trip to the California border was a once in a lifetime, but every other weekend I was doing something with my bike. Two of my friends, Doug Fraser and Freddie Baird, were also members of the Vancouver Motorcycle Club. We all had Harleys, and some weekends we'd go into the States or go up country into the interior toward Jasper and stay over for a night. One time we rode to the Grand Coulee Dam in Washington, and on the way back we stopped at Leavenworth to get a motel for the weekend. We arrived too late; the motels' "No Vacancy" signs were lit up and they were all booked full. I got the bright idea of going to the police station to see if they could put us up for the night. Doug and Freddie weren't as convinced this was a good plan, but they said it was worth a try.

I walked into the station and saw the jailer sitting at his desk. I introduced myself and said there were three of us riding bikes. His response surprised me.

"I know you're here. We've been watching you come through."

I pointed to the vacant cell and asked if he could put us up for the night. He moved slowly, like he wasn't crazy about it, and unlocked the cell door. He didn't say anything, just waved us in. The cell had benches, so we each stretched out and enjoyed the warmth of the place. We could hear four or five cells were occupied, but we felt safe, especially when the jailer left the door unlocked. In the morning, he walked past our cell with a big pot of steaming hot coffee. We asked if we could have some. He glared at us and said, "Get the heck out. If

Stuck in the mud on the BSA bike.

my chief catches you here, he'll have my hide." We took his advice and headed home.

I bought a smaller English motorcycle, a BSA that was good for mud racing. I had to coat the wires to waterproof them. We'd ride in a field in Burnaby. Trevor Deeley, the son of the owner of the motorcycle dealership on Broadway, used to race with us. We'd start out clean, but by the end of the race we'd all be splattered with mud. I had a good time trying to keep the bike steady through the ruts and tracks. Didn't much care about getting dirty.

Life went on pretty much the same until a possibility for work came up in Fort Langley. MacMillan Bloedel bought the McDonald brothers' mill, so they pulled up stakes and went over to Fort Langley. I stayed on and worked for the new company for a while, but Cam went with Bobby and Ian. The brothers had bought a burnt-out mill in Fort Langley; it had burned right to the ground. Fire was a real hazard with so much wood on mill property lying around, which

The lumber carrier has aged over the years but still looks good to me today.

is why no one was allowed to smoke. I smoked a little until I was twenty-seven, and I remember seeing a couple of guys going into the washroom for their smokes. It wasn't allowed so you had to be careful not to get caught. Nobody wanted to see a mill catch fire.

The one in Fort Langley was on a good piece of property for a mill. The river was right there. In 1953, Cam called me again and said, "I need you here to help me build the mill." I quit my job in New West and went over to Fort Langley to clean up the yard and hire guys to help build the mill, which they were calling McDonald Cedar Products.

While doing the work, Cam told me he had purchased a lumber carrier at a used machinery company in the industrial area of Vancouver. He needed my help to get it to the mill. He took me to the used equipment dealer's warehouse to pick it up. The plan was that I would drive it back to the mill to save on the costs of loading it on a truck. The carrier had four wheels, a steering wheel, gearshift and a windshield, but it could only travel at about thirty miles per hour, an adequate speed around the mill yard but not the best for highway driving.

This was a large, high machine that straddled a load of lumber that came off the mill's green chain. The machine worked well when the load was about four feet high, slipped lengthwise front to back on the carrier block under the driver. Without the lumber, the empty space made it easy to walk through if you bent over a little. Cam explained I was going to have to drive several miles down city main streets and portions of the highway and across a bridge before we would arrive at the mill.

I climbed up about six feet on the ladder on the driver's side and settled into the seat that raised me a couple more feet. I followed Cam along the streets, driving below the speed limit. People don't usually see this kind of machine, especially on the highway. Cars avoided me by giving me a wide berth as they passed. Several cars honked to warn me they were near or just to say hello. Some passengers rolled down their windows and gave me funny looks. Fortunately, the day was clear and not raining, so I could see what was coming. I ignored the honks and stares and kept focused on the road. I'd driven these machines in mills, but never out on public roads, but I felt pretty secure because Cam stayed close up front and would wait for me if I got too far back.

The journey took about an hour and a half. I followed Cam through Vancouver to New Westminster, across the Pattullo Bridge, over to Whalley, and worked my way to Fort Langley. The view from up high was impressive in spots as I rumbled along, shifting down to low gear at the stop signs and turning the big steering wheel as I manoeuvred around corners. At intersections, I was a bit worried that an Austin Mini or a Volkswagen Bug might drive underneath and get stuck, but traffic was light so we got to the mill without any accident or problems. That was the only time I drove a lumber carrier down city streets, and I'm probably the only person who ever made that small trek along a highway and over a bridge. I was glad to park it, climb down and get back to my regular work.

I stayed at McDonald Cedar for about five years. A year after I got the night shift underway, the McDonald brothers moved me to the day shift where I became the assistant superintendent to Cam. The McDonalds liked him because he knew the siding business and he had helped get the Fort Langley mill off the ground, but he could be tough.

Cam was several years older than me, and he'd helped me get set up a few times in the mills. We were friends, and I learned a lot from him. He knew I knew cedar siding as well as he did and that I'd do a good job for him, so he didn't give me any problems as a boss, but I could see he was a hard taskmaster with the guys. I got along good with the guys because I'd interviewed them and hired most of them. I learned early on that a boss doesn't need to swear or yell at employees to get the best work performance. Treat them fair and be polite, and they'll do a good job. Most of the guys lived around Fort Langley, had houses there, so they appreciated having a stable job close to home.

One day, we had some excitement at the mill. Bobby McDonald liked the professional hockey player Babe Pratt. In the 1930s and '40s, he'd been a defenceman for teams like the New York Rangers and the Toronto Maple Leafs. By 1953, he had played three years for the New Westminster Royals in the Western Hockey League and then ended his career playing for the Tacoma Rockets. Babe was looking for work, so Bobby hired him to come in on the night shift.

The men thought it was great to have a guy like Babe with his record working with them. He was like a hero to them. He'd scored the winning goal in a Stanley Cup final and was inducted into the Hockey Hall of Fame. He'd learned to skate when he was around six or seven and was on a championship peewee team in Winnipeg in 1926. He turned pro in 1935 and played with the Rangers. I once asked him how he got the name Babe, and he said when he was a kid he'd gone to play baseball with his older brother, and he'd hit the ball into the infield a couple of times. "Somebody said, 'he's a regular Babe Ruth,' and the next day everybody was calling me Babe. Nobody called me Walter." Sounded familiar. Nobody calls me by my given name either.

After I'd been working at McDonald Cedar for a while, something memorable happened. The day was hot, so Cam said, "Let's go for something cold to drink." We walked over to a small café in Fort Langley that had a deep freeze and a couple of stools off to the side of the main café. Cam said, "How about an ice cream cone?" I said sure. As the girl behind the counter took our orders for two cones, I couldn't help notice she was very pretty. I'd never been there before and noticed they had a collection of antique artifacts on the walls. I was looking at them when I felt Cam nudge my arm. I turned to see what he wanted, and he nodded his head toward the deep freeze. The ice cream must have been way at the bottom because the girl was leaning into the freezer. Cam was looking at her legs. When she handed us the cones, she made an impression on me, and I decided to return the next day without Cam.

She was there working again when I walked in for another ice cream cone. We got talking, and she told me her name was Marilyn. She seemed more grown up than her eighteen years, and she was beautiful, so I asked her if she'd like to go for a ride later. I had a bright red 1953 Ford convertible with the continental kit—the spare tire on the back for design, so I knew she'd like the car. We decided to go to the Langley drive-in. We ended up talking the whole time, so I asked her out again and we started dating. I didn't normally go out much because I was working, but even then, I knew she was the one.

A couple of weeks after this, around 2:00 in the morning when I was getting off the night shift, I saw a young lady sitting in

Marilyn's Grade 10 school photo.

my car. She was the daughter of the guy who owned the local garage. I asked her what she was doing there, and she said she wanted to get to know me better. I was interested in Marilyn, so I told her the best I could do was drive her home. I think she liked my car.

In the summer of 1954, I decided to ask Marilyn to marry me. I drove into New Westminster and bought an engagement ring. We'd been getting along really well, so I was confident she'd say yes. On the evening of August 16, we were sitting in the car and I showed her the ring. I said, "Will you marry me?" She said, "Yes." Simple as that, she made me the happiest man in town.

The second item I bought for Marilyn was a winter coat. She didn't own one, and I knew we'd be getting into the cold months pretty soon, so I took her to New Westminster where she found a warm coat with a fur collar. Her family had done the best they could for her, but money was tight.

LEFT Marilyn and her dad Mike.

RIGHT John and Eudokia Czorny. This photo was taken a few years after they arrived in BC.

Marilyn was born in 1936 in a farmhouse about three miles northeast of Glendon, Alberta, but she didn't live there long. One of the first things I learned about her family was they were always on the move because of her dad's work.

The Czorny family, which included Marilyn's older brother Sandy, her mom Nancy and her dad Mike, stayed only for a short time in spots between Alberta and British Columbia. Mike worked as a section man for the Canadian National Railway, and the company was always sending him off to different parts of the West to be in charge of sections of the rail line. Wherever he ended up, he made sure his crew kept his two-hundred-mile line of track clean. The family knew he could be transferred at any time and they kind of got used to it.

Mike was an interesting guy and we got along from the start. He was born in 1911 in Tarnopol, Galicia, Poland. He immigrated to Canada in 1929 on the ship *Regina* two years after his father John

Mike (wearing the overalls) with his crew.

had arrived here. Mike's mother Julia and his brother and three sisters were still back in Poland. Mike started working in BC for the CNR the same year he landed, and when he got laid off he travelled all over western Canada looking for farm work. In the 1930s, men took work where they could find it. One time, when he was riding the rails, he got locked inside a boxcar that was set out on a siding track between Calgary and Edmonton. The boxcar sat there for five days, and Mike thought he was a goner until a passing trainman heard him calling for help. That experience didn't scare him off trains because he landed more work as a section man. Both Mike and Nancy came from hard-working and loving families. I saw early on that they passed these virtues on to Marilyn and her brother Sandy.

Marilyn's grandparents were from Poland and Russia. Mike's father, John Czorny, also worked for the CNR, and after Mike's mother Julia died, John married Eudokia Riznyk and moved to Lucerne, BC.

Nancy's side of the family came from Russia. Her father, Jacob Chibanoff, was born in 1881 in the city of Archangel'sk in northwest Russia on the White Sea, where he served as a sailor, second degree, on the famous battleship *Potemkin*. Marilyn's grandfather told her the story of his adventure on the ship: in 1905, he and his shipmates found the meat in their borscht was full of maggots. One of the seamen complained to their commander, and the commander got so angry that he took out his gun and shot the man. He killed the guy in front of a few other seamen, so they grabbed him and threw him overboard. Then they turned on the other officers and killed them. The ship ended up in a Russian port where workers had been on strike. When the people heard about the sailors' bad treatment, they had more protests that helped kick off the Russian Revolution.

Nancy's mother, Maria Evanoff, also from Russia, came from an area east of the Black Sea. Her life seemed pretty normal. Maria married Jacob (now John) Chibanoff. They arrived in Halifax in 1911 and after a few months moved to a Doukhobor community in southeastern BC. They soon left that area and settled in Mortonmoor, Alberta, where Marilyn's mother was born in 1916. Six of their children died at birth or soon after. Marilyn's grandfather John died in 1949, and her grandmother Maria lived with them until she died in 1951. Both are buried in Glendon, Alberta, the small town three miles from where Sandy and Marilyn were born.

In those days, women had their children at home. Grandmother Maria was the midwife and helped deliver a baby boy. Nancy wanted to name her son after her brother Alexander Chibanoff, who everyone called Sandy. Years later, during World War II, her brother was in service overseas as a Royal Canadian Air Force pilot, where he flew about three hours every day in two- and three-engine planes. They would go out for air-to-sea training and practise bombing missions, except when the haze got thick and visibility was bad.

This routine went on for some time until two weeks after he wrote Nancy his last letter in September of '43; he was killed in action near Middlesex, England.

Nancy learned that two boys had witnessed her brother's crash. George Rust and his twin brother were playing close to home when they saw Sandy's Halifax bomber smash into a field on their family farm. We found out much later that, every year on Remembrance Day, George and his wife Rosemary laid a wreath for Marilyn's Uncle Sandy and his crew, which is why, when we travelled to England, we made a point of meeting them and visiting the site.

While they were still living on the Glendon farm, Marilyn and her brother Sandy often stayed with the Chibanoff grandparents to help out. They were hard-working kids, but they had fun too. In the summer, they picked up rocks to clear the fields for cattle grazing, and they took sled rides and went snowshoeing in winter. They even used frozen cow turds as makeshift toboggans. Her grandfather looked after the farm while her grandmother worked in the house. Marilyn remembered her grandmother Maria as a physically strong woman: "she could carry pails of water from the well and hook up the team of horses to go on her own into town for supplies." She was also a good cook and kept a vegetable garden. She took Marilyn berry picking and showed her how to can fruit and bake pies. Marilyn and Sandy always had chores, which was part of being a family—everyone pitched in and felt good about it. She grew up believing if you work hard, life will become easier.

From Glendon, Marilyn and her family headed to Red Pass in BC, where Marilyn's parents had met. In 1934, Nancy went to live with her sister Mary nearby in Rainbow, BC. She found work as a waitress in Red Pass Junction, six miles west of Rainbow. The Red Pass Junction station of the CN Railway was originally built as a station at the junction of the Grand Truck Pacific and Canadian Northern

Mike and Nancy at their house on Glover Road in Fort Langley in the early 1950s.

Railway lines between Tête Jaune Cache and Jasper, Alberta. This junction closed fifty years later. At the time, Red Pass had a hotel, general store, railway station, three section houses and about four private homes. While working as a waitress, Nancy met a handsome guy called Mike Czorny. Every Saturday he hopped a train or drove a railroad motor car on the rail line to travel the twenty-two miles from Jackman Flats to Red Pass to see her. In 1935 they sealed the deal and were married.

When Marilyn was growing up, no one had time to talk about the past. She once told me, "I wish I knew more about our early family life. I wish Mom and Dad had shared more stories by the light of the kerosene lamp or candlelight. Time was and is so precious to communicate and share our feelings and thoughts, talk about the

LEFT Mike, Nancy, Eudokia and John.

RIGHT Mike and Nancy in back. Sandy and Marilyn on grandpa John's lap.

past and learn about the hardships and the good times." But no one had the time. In those days, everyone was too busy working and trying to survive. Most people were too tired to sit around and tell those stories.

Marilyn and her family stayed in Red Pass long enough to have a vegetable garden and to keep a few chickens, which helped them survive the Depression. They soon moved again and lived about twenty miles outside of Jasper in Lucerne, BC, where Mike's father and stepmother lived.

Their time in Lucerne was spent trying to keep afloat financially. Mike hunted and also fished in Lucerne Lake while Nancy bought fresh fruit from the local grocery store for a sweet treat. Sugar was rationed and could only be bought with government-issued coupons. During those lean years, Marilyn learned to eat nutritious food and to watch her pennies; any money she earned picking fruit and mowing lawns went back into the family coffers.

Between Jasper, Alberta, and Blue River, BC—a distance of about 130 miles—the Czorny family moved about twenty times. They lived in Lucerne five times. During the war, Marilyn and Sandy were living near one of the Japanese internment camps and, kids being kids, played ball with the camp children. That was probably the same camp I saw when I rode the train to Vancouver from Winnipeg during the war.

Some of the places where the Czorny family lived were pretty basic. In a couple of cases, their home was a railway section house close to a railway track where trains rumbled past their front door every day. All this moving around made it hard for Marilyn and her brother to stay in a regular school for the whole year. Some towns were home for only a few months before Mike got transferred and they pulled up stakes again, so they took correspondence courses to keep up with their schoolwork. When they did have the chance to go to school, they walked a couple of miles to the nearest one-room schoolhouse. One morning in winter, the temperature hit −58°F. That day, when they started out for school, they stopped at every neighbour's house along the route to keep warm. They got about half a mile from home before they decided to turn back and call it a day. Mike kept getting bumped from each place because he didn't have enough seniority to lay down roots, but a job was a job and he was happy to be working.

When Marilyn was about ten years old, they were living in the Fraser Valley in a small town called Atchelitz, about seven miles southwest of Chilliwack. One time, she and three friends took the BC Electric Railway into Chilliwack to see a movie, but they got separated at the end of the show when Marilyn went to buy an ice cream cone. They caught the train back, but Marilyn missed it, so she had to walk the seven miles home. Her friends got heck for leaving her behind, but she explained to them that ice cream always had a powerful pull on her stomach because she'd loved it from a young age. Her dad would sometimes bring home a tub of it for her and Sandy. What made the treat so special was that on his route home, her dad would stand outside at the back of the train in the bitter cold to keep the ice cream from melting along the way.

In 1947, Mike got work in Fort Langley where Marilyn started elementary school. She always said she found school a little difficult because she never properly learned her phonics, and she felt self-conscious as a tall, lanky girl among younger and smaller students. In grade six, her teacher gave her some much-needed encouragement, which boosted her confidence. The next year, she was part of the group of first students to attend grades seven to twelve in the newly built Langley Senior High School. Her dad was away for long stretches of time and came home only once a month, so her mom had to stay home and keep an eye on the kids. Her mother rarely went out except to shop in Fort Langley at Vickerie's grocery store.

"They were so good to us," Marilyn told me. "We ran up a bill and we could only pay a little each month when we got a cheque from Dad. They trusted us and kept us fed for years with the bill outstanding. I knew Mom was concerned about our finances. In the summer, Sandy and I would pick strawberries, raspberries and blackberries for a little money. We'd pick hops and sell them to a hops company that made beer in Chilliwack. We also fished oolichans. You name it; we tried it to help out to pay up that bill at Vickerie's."

While Marilyn's brother was still in high school, he worked weekends at McDonald Cedar Products in Fort Langley where I

I gave Marilyn this photo before she moved to Smithers to remind her that I loved her.

happened to be working as foreman. I had no idea that we would end up being brothers-in-law.

When Marilyn and I started dating and during our engagement, Sandy and her parents were still living in Fort Langley but in 1954, Mike, Nancy and Sandy moved to Smithers, BC, where Mike got a job as road master on the Telkwa subdivision. The other important event was Marilyn's baby sister Charlene, who was born in February of that year. The family lived in Smithers until 1962, when Mike got promoted to supervisor of track in Edmonton.

Marilyn stayed in Fort Langley to graduate from her high school in 1955. She lived with Cam, his wife Phyllis and their children during her Grade 12 year and worked at the Fort Café to earn some spending money. I had gone to a professional photographer to have my picture taken for her, which turned out to be a good idea because she was thinking about moving to Smithers for a while: I hoped the picture would remind her that I loved her.

Marilyn was missing her parents and her brother and she wanted to be sure we were doing the right thing getting married, so before we started making wedding plans, she moved to Smithers for six months. We sometimes communicated through phone, but long distance was expensive, so we sent letters. We had a system: Marilyn was working for the CNR in Smithers doing secretarial work. She'd put her letters in an envelope and send it down from her train office to the one in Fort Langley. At McDonald Cedar, we'd be shipping lumber on trucks or in boxcars out of Fort Langley, and I'd walk over to the CNR office and drop off the shipping papers. The guy in the office would tell me he had a delivery for me and hand me the envelope with Marilyn's letters. She sent more to me than I sent to Smithers, but I used the same system. We were making wedding plans while she was in Smithers and talking about where we'd live when, suddenly, a bit of good fortune came our way.

4

HIRED FIRED HIRED

A STROKE OF LUCK came my way when Ian McDonald brought me a letter from a friend of his called Ormie Harris. Ormie lived in West Vancouver and had a sawmill near Kamloops, but he also had a summer house in Fort Langley that was on a 160-acre farm. A little ways from the big main house was a two-bedroom caretaker's house, and it was empty. Ormie was writing to Ian to ask if he knew anyone who might be interested in caretaking a farm rent-free. The only condition was that the man had to be married.

I said to Ian, "I'm not married yet."

"You're going to be."

"But not for a year."

Ian suggested I go see him and take the letter as an introduction. On the weekend, I drove over to the farm with the letter and met Ormie. I explained that I wouldn't be married for a few months, but I wouldn't mind looking after his farm. He figured I'd be a good caretaker, so we agreed I would move in right away. The farm had forty head of cattle. In the winter I would feed the steers and in the summer I'd mow the yard, look after the swimming pool, and do the haying. Ormie came out on the weekends, but I didn't need to go near the big house.

At the time, I had been boarding five minutes away from the farm at a house on Wright Road with an elderly lady, Mrs. Wall, and her husband. As soon as I got the keys to the caretaker's house, I moved out, but Mrs. Wall offered to keep making my meals, so I paid her for my food. She made a lunch kit for me to take to work at McDonald Cedar. I was on afternoon shift from 4:30 to 2:00 a.m., so I'd eat my breakfast at the caretaker's house, pick up my lunch, go home and stop in at her place for dinner before my shift started.

During that time, I flew up to Smithers twice to see Marilyn and to iron out the details for our wedding. I'd fly up Friday night and then visit for the whole weekend. We had some good times, especially when her brother Sandy and his friends would have a jam session and play their music well past midnight. Their mom Nancy would sit with us and tap her foot and clap along. I could see she was proud of her kids.

When I got back to Fort Langley, I started getting the house ready for Marilyn. Our wedding day was May 19, 1956, two days after Marilyn's twentieth birthday. I was twenty-seven. We didn't expect our parents to pay, so we made it a simple gathering that we could afford. We were green at making plans, but the day worked out okay except for an unforeseen conflict in dates.

Sandy had been working on the railroad in Smithers, but he'd decided he wanted to join the RCMP. The police sent him to Prince Rupert for tests and accepted him. He would be sent to Regina or Ottawa to start his training in April. When he saw that the start of the course coincided with our May 19 date, he asked if he could take a week off from training to attend his sister's wedding. His superior told him that once training started, he would be confined to the

LEFT Marilyn and me on our wedding day, May 19, 1956.

RIGHT Our wedding party. Left to right: Marilyn's parents Nancy and Mike, two bridesmaids, Gerry (the maid of honour, who was married two weeks later and became Gerry Carlson; Marilyn was her matron of honour), the bride and groom, my best man Walt Cleave, my parents Florence and William.

base for the entire eight weeks, so no, they told him, he would not be able to leave. Sandy thought about this and figured his sister's wedding was more important to him, so he cancelled his training with the RCMP and chose to stay on at the railroad. Marilyn was particularly glad to see her brother there on her big day. We were married in the Fort Langley Anglican Church and then drove to San Francisco for our honeymoon.

We lived in that caretaker's house rent-free for six years and had our first daughter Wendy there. When Marilyn got pregnant the second time, the outcome was not good: our next baby girl was stillborn. I had taken Marilyn to the hospital in Langley like I had the time before with Wendy. I thought everything would go the same and I'd be back later to check in on them. They didn't let dads in the labour room, so I said I'd see her in a little while and went home. Soon I had a phone call, a sad call, from the hospital. They hardly kept Marilyn in the hospital, only for a couple of days. That was rough.

Marilyn and I mourned the loss of our second baby. We never gave her a name, but we did have a burial in a plot we had bought in a cemetery close to our home. Of course it was sad for us, but somehow we dealt with it. You had to. Life goes on. We were still living on the farm when a short time later our daughter Suzanne came along.

One summer, while working at McDonald Cedar, Bob's son David came to work at his dad's mill. He had brought his friend from high school, John Chipperfield (who later became a lawyer in New Westminster), to earn a little money over the vacation. I was their foreman and showed them how to make pickets. They were both good workers.

That same summer, Marilyn and I invited David over for a swim at the Harris farm. We were sitting on the pool deck watching him jump off the diving board and would cheer each time he climbed back up to do a few fancy dives, but we stopped cheering when he sprang up really high into the air and landed on the diving board so hard it snapped in two. He fell into the pool and fortunately wasn't hurt, but he felt terrible that he'd broken Mr. Harris's diving board.

We hauled out the broken piece and stood staring at it.

"Can we fix it?" David asked.

"No."

I scratched my head for a minute. "We need a new diving board."

"I'll pay for it," he offered. "Where do you buy them?"

"I don't know."

Marilyn joined us and said, "Chick, it's a board. A piece of wood."

Only took a couple of seconds for the penny to drop. "The mill!"

I drove over to the mill and asked one of the workers to cut out a good plank. The millwright got a wrench and drilled some holes, and I painted it and slid the new board into place. I don't think Ormie Harris ever knew it had been broken.

David turned out to be a successful businessman. He and his brother Stephen formed Western Cablevision that served New Westminster and Surrey, and for almost thirty years, David and his wife Joanne have been donors to the Royal Columbian Hospital Foundation in New Westminster.

On another memorable day, Cam Richie told me he'd give me a free box of twelve shells for every pheasant I shot for him. I figured this would be easy work because the fields were full of pheasants as well as ducks, geese and other wild birds. We also had a lake full of ducks, so I asked Marilyn if she'd like to learn how to shoot. She thought this was a good idea, so I gave her a loaded double barrel 16-gauge shotgun and together we went out along an old gravel road to the lake to shoot ducks. I had a pump 12-gauge with three shells, and I fired and got a couple. A few minutes later, the birds flew up and I heard a bullet whiz past my ear.

For some reason, I had forgotten to tell her to never aim the gun at anyone. I turned around to see Marilyn behind me with her eyes focused on the sky but her gun aimed straight ahead. We took time that day to iron out the kinks in her method. A few days later I came home from lunch and saw a dead pheasant hanging from the clothesline.

"Where'd that come from?" I asked.

"I shot it."

Marilyn was a quick learner and after that was always careful how she handled a gun.

In the 1970s, Bob and Ian sold their holdings to the company now known as Interfor. After Bob passed away, Ian and David put together an annual company party at the Vancouver Golf Club in Coquitlam for the ex-senior employees and customers. Every Christmas season, several of us would get together for dinner and drinks. Of the dozen guys who attended, I remember Paul McCracken from Tumac Lumber in Portland Oregon; Mike LePage from Barnett Lumber in Vancouver; Cliff Glover, the head of Cedar Sales at Seaboard Lumber and Export Group in Vancouver; Jack Amm, a log manager at McDonald Cedar; Ted Smaback, a mill foreman; and Ken Bradley, a salesman at MacDonald Cedar, now in his nineties. In 2016, I attended a lunch with David, Mike, Cliff, Jack, Ken and their secretary Marilyn. We always stay in touch.

After six years of living at Mr. Harris's place and working in Langley, I decided it was time for a change. I could see I wasn't really advancing in that particular job and felt I could go further if I tried something else. I saw an ad in the newspaper for a sawmill job up the coast in Tahsis on Vancouver Island and told Marilyn I was going to apply for it. Not long after I applied, they called me in for an interview.

"Yes, we'll hire you, and we'll move your furniture up to the town. No expense to you."

Marilyn was happy we could take our things with us. The movers loaded our furniture into two big containers and barged them up the coast, which took a few days. A moving van then moved everything to our new home in Tahsis. For the first couple of days, before the company house was ready, we stayed at a hotel. Wendy was five years old and Suzanne was only four months.

Tahsis is a village in Nootka Sound about 185 miles northwest of Victoria. In the 1930s, a number of lumber companies tried to open sawmills on the west coast of Vancouver Island. In its heyday there were about twenty-five hundred people living in Tahsis. The area was a good place to build a sawmill: the land was flat at the mouth of an inlet and had deep-sea access for ships. The mill faced southeast, getting enough sunlight to protect the fresh-cut lumber from mould, and this was an advantage because it could rain for days with no let-up. The rain and the isolation, though, were two things that made it a tough place for a young family.

Marilyn and I felt like Tahsis was in no man's land. There were no roads into the town, so you could only get in by boat or floatplane. I had a supervisor's job as yard foreman, and I knew I could do the job because I'd worked as yard foreman at McDonald Cedar. I was moving up in the industry, but the place was remote. Nonetheless, we tried to make the best of things. Marilyn made friends with the wives of the supervisors and some of those friendships lasted many years. Plus, we got to know our neighbours, which helped us feel part of the community.

A German doctor lived right next door to us, and it was a good thing since a couple of days after we arrived Suzanne got really sick with a bad cold. We went next door and asked her what we should do. She came over, took a look at Suzanne, and said, "She's got a fever. Do you have a washtub?" We said yes. She told us to fill it with cold water, which we did. We got kind of nervous when she gently placed the baby in it. Suzanne was screaming, and we hoped the doctor knew what she was doing, but we didn't want to question her method; we wanted to trust her, but we both had to stop ourselves from reaching down and pulling Suzanne out. Good thing we didn't. The doctor brought Suzanne out of the water after only a few seconds, and that sudden jolt of cold brought the fever down right away.

Even though I was at work all day, my job never kept me away from my family, and Marilyn knew to call me if she needed me.

My job kept me busy and gave me more experience. I was yard foreman responsible for thirty acres of land, and was in charge of the guys lining up the lumber to be loaded on ships destined for Europe, places like England. Custom-ordered lumber was also being shipped to Japan. It took three to four days to load all of the lumber on those huge ships. Fifteen days later a new ship would come in. I had to keep track of all the wood that came into the yard, and I was in charge of the lumber coming out of the mill for that day. I would number each order because I wanted everything to be precise and in the right place. The shippers would look at my line-up and find I'd planned the whole yard out. Let's say the wood they wanted was in Bay 2, Alley 5. When the ship came in, the shipper would come to me and ask, "Where's the wood for that ship?" I knew right away to go to Bay 2, Alley 5. We had few problems; my system worked really well.

Soon after we got there, I was alone for a while. Marilyn's brother Sandy was getting married to June, and Marilyn flew out to the wedding with Wendy and Suzanne. I asked the mill manager if I could have a couple of days off to go to my brother-in-law's wedding. He didn't look too happy. I understood why when he said, "If I let you go, do you plan to come back?" He was afraid I'd leave for good. A lot of workers would leave town and not return because they couldn't take the isolation. I wouldn't have gone away and left all of our furniture, but the manager didn't know that. In the end, I did go to Sandy and June's wedding, but I came back alone. Marilyn and the girls stayed a few days longer.

I was lonely during their absence, but I kept working. I knew it was a temporary job, and I was gaining experience, so a few months there wouldn't kill me. I took advantage of my time in Tahsis to learn and to take classes. After you've been in the mills awhile, you can take grading classes to become a certified grader or you can write the tallying exam.

In Tahsis I met two fellows, Gord and Roy, who like me, were keen to write the tallying exam. New guys never took that exam, but we'd been in the business long enough to qualify. The exam would test us on things like calculating board footage—the average width, thickness and length of an order or tally; how to figure values and correct tally construction sizes; and how to calculate boxcar capacity and estimate shipment weights.

The exam usually took three hours, but we did it in an hour and a half. We came out of the exam with all of our totals. The first question on the exam was, "How many pieces of wood that are all the same size go into a lumber car?" Roy and I had the right answer. Gord had made a small miscalculation that messed up his total in number one. I can still hear him complaining about missing that first question. He got 91 percent and Roy and I got 100 percent.

Gord's story has a happy ending. He had a very good sense of smell, which is an asset in the mill, especially if you're a grader: he could tell the difference between fir, cedar, hemlock and cypress. He decided to become a certified grader and was doing so well in his grading classes that he got hired in England as a timber inspector. Marilyn and I went over there to visit, and we had a great time touring England with Gord and his wife. He did okay for himself even though he messed up question one.

We stayed in Tahsis for about eleven months. It doesn't sound like a long time, but that was the kind of deal I made. Like I said, Tahsis was isolated, but another problem was it'd get a lot of transients, not the kind of place to raise a family. After several months, I said, "Okay, this is enough." Marilyn agreed. But I didn't leave without having some place to go to. The movers loaded up all of our stuff and brought it to a house we rented in the Como Lake area in Coquitlam. I'd applied for a job with the Nalos Lumber Company, a cedar mill located on CPR leased land in False Creek on what became the Expo 86 lands. I was hired in their mill as the night-shift foreman.

In the short time I was there, I made an impression on the boss, but not the kind you might think. We would be cutting cedar logs on the night shift while another foreman ran the day shift. After being there only a few months, I thought we were doing okay. The tally sheets from each shift showed how many logs came in and how much wood we cut, and each day I could see we were doing as well as the day shift, and sometimes even out-cutting them. The day shift, however, is supposed to be the most productive; most night shifts are just expected to keep the machinery going.

Early one morning after my shift ended, the superintendent who was in charge of both shifts called me over. "Chick," he said, "Mr. Nalos wants to see you."

I had no idea what to expect, but never imagined the worst. I went up to Mr. Nalos's office, where he was sitting behind his desk with another man standing beside him—his key office guy, I guess. He never asked me to sit down. After a short greeting, he said, "Chick, it's not working out. The men are angry. You're working them too hard and it's bad for morale."

"They're angry?" I was stumped by this. No one had complained to me.

"That's right," he said. "I'm sorry, but we're going to have to let you go."

He wasn't angry or anything, but he didn't give me a chance to tell my story. I had only been working there for a few months, so they really didn't know me. Mr. Nalos believed the report he'd received from his head guy and that was final. He handed me my cheque and I left. I had no idea when I went to his office that he was going to fire me because I thought we were doing a good job, but that was the end of it, no two weeks notice, just "here's your cheque, you're finished." I never got a chance to go back into the mill and talk to the guys. They paid me for my work right up to the time I stood in his office, so I knew they were ready to let me go when they called me in. I couldn't believe it.

Something wasn't right. I knew my crew wasn't complaining; the men were all working well together. The night shift was working steady—we had a really good sawyer who knew what he was doing, and he could keep the saw going. In a mill you could have a fast guy or someone plugging up the works, but we never stopped the head end. We knew we were faster than the day shift, so maybe that didn't sit well with the boss on the day shift.

The Nalos Lumber Mill was the first and only company that ever fired me. In all the years working in mills, no one had ever told me I wasn't working hard or I was doing a bad job. I'd learned how to get the best work out of my guys: they were hard workers and they kept everything running smoothly, without me ever swearing at them or driving them too hard. But here Mr. Nalos was telling me I was bad for morale.

A few years later, one of the guys from the Nalos day shift was out our way in Port Kells and he dropped by to talk. He told me the reason they fired me was the superintendent was embarrassed that the night shift was out-cutting the day shift. I wish I'd known that when I was driving home in the early hours of that morning.

That day when I got fired, I was driving slower than usual on my way home. I was disappointed and kept thinking about what I was going to tell Marilyn. We had just moved from Tahsis and were try-ing to make a go of it, and now I didn't have a job. Normally, when I'd get home around 2:00 in the morning, Marilyn and I would talk about the work, and she knew I was doing my best. As I drove, I thought about Tahsis, and how the manager wanted me to stay. Had I made a mistake in leaving there? Another thought popped in that said, "No, Tahsis wasn't the place to raise a family." The move away from there was right, but how would I tell my wife I'd been discharged? Fortunately, Marilyn understood the situation and reassured me that something else would come up. Her encouragement was nice to hear, and I didn't have time to lick my wounds: I had a family to support.

Luckily, I soon found work. I saw a job advertised for Capilano Cedar Timber, a company owned by the McClelland family in New Westminster. Bill Robinson was working as their manager looking after sales. They didn't own any sawmills but contracted out to three mills for companies needing custom-cut lumber. Bill Robinson took me around to the mills and introduced me so they'd know I'd be coming around. The three sawmills cutting for McClelland were scattered through the Lower Mainland: Ranjit Mattu was running Hem-Fir (for hemlock and fir) in Vancouver along the Fraser River and near the airport; a guy named Munro was in charge of one in North Vancouver; and Harold Morrison owned one in Port Kells. The McClellands worked out of their New Westminster office, but my job was on the road: I had to drive to the three locations, make sure the orders were in and check on the crew.

I was used to working in one place pretty much, but this job was all about travelling. I had to drive several miles in a triangle to the three mills. I have to admit I didn't care for it, but the company was cutting different kinds of lumber, and I was learning something new, so I kept at it for a year. This was when I met Harold Morrison and a new chapter started in my life.

5

"YOU HAVE TO START SOMEWHERE"

AROLD MORRISON owned the sawmill in Port Kells, one of the mills I visited for Capilano Cedar Timber. He and his wife had an apartment above the office, so he lived and worked at the mill. He was about sixty years old and had been in the mill business for a long while when he told me he wanted to get out of it.

"Chick, I've been taking time off for vacations these past couple of years, and every time I leave, the crew gets time off even though they want to work. They need permanent work and I want more vacations, so I'm going to sell the business."

Harold was still getting orders, but his practice of closing down the mill to go away for a couple of weeks was losing him money and customers. His millwright, Emil Geering, lived on the property and would keep an eye on the place. He stayed working when the crew, mostly farmers, went "on vacation." The mill was steam-run, and Emil was the only guy who knew how to operate it, but they also had to have a steam engineer on site with a ticket to run steam boilers.

I looked at the mill, at Harold and at Emil, and the wheels in my head started turning. I had played with the idea of being my own boss, but didn't really have a plan to buy a mill at that time, but I started thinking, "Maybe I should buy this one."

My family had been poor. I saw no future if I had wanted to become a doctor or a lawyer because that would have meant being prepared to pay a large sum of money for my education. I decided to enter the sawmill business because it was all I knew from my teenage and working years and I wanted to earn my own future, and now I saw the opportunity to make a go of it.

It wasn't a bad little mill with its circular saws and steam, and I knew from other mills I'd worked at that I could change the steam over to electricity. But this purchase was too big a deal. We didn't have any money, and Marilyn figured it was out of the question, but she could see I kept mulling it over. I had been a foreman at the age of twenty-one at B.C. Manufacturing. That was pretty young. A lot of guys wanted their own sawmill, and I knew opportunities like this didn't come by often. I told her it was a stroke of luck that I was working at Capilano at the same time Harold wanted to sell, so maybe we should find a way to give it a go. Marilyn suggested I talk to her brother Sandy for his advice. Sandy said, "You have to start somewhere." He suggested we borrow money from their aunt to help us out.

The business would cost around $60,000 for two acres of land east of the mill and adjoining the Fraser River, as well as for the custom-cutting mill. Emil was the heart of the operation, so I knew we'd want to keep him on. I say *we* because I couldn't do this alone.

I started considering who would make a good partner. I knew Vic Rempel was a really good sawyer and that together we just might do this. We were seventeen when we met at B.C. Manufacturing in New Westminster, which was where he still lived.

My decision to ask Vic was based partly on his work background, but he and I had been friends for years. (He's in the photo with Willie,

Phil and me on the 1947 Harleys.) He had a great disposition and common sense. Even though he only went to Grade 11, he was smart, especially with money. He was also an honest guy who wasn't afraid of hard work.

Vic and I had become friends over our shared love of motorcycles. When we were still teenagers, money was tight. I had saved enough to buy my Harley army bike for $365, and Vic went in half with Dick Klein to share a bike. We'd hang around on the weekends and our friends would call Vic, Dick and me the Ick Brothers. Later, my best man Walt Cleave got Vic to buy a dirt bike, and he loved the sport, especially the competition. He won trophies throughout his life, right into the old-timers class for guys over forty. If he wasn't racing at Mission, Aldergrove and Agassiz every weekend, he'd be tuning motorcycles or building engines.

Vic had been an athletic guy all his life. He did gymnastics and, starting when he was a kid, he liked to swim and dive at the YMCA in New Westminster. He also played basketball in high school. In his fifties, he told me he'd gained only five pounds since those high school days. On Friday nights in the summer, after we'd been at a dance in Maple Ridge or Pitt Meadows, sometimes we'd drive to a section of the Alouette River where a small, natural pool was perfect for swimming. We'd get there around 2:00 or 3:00 in the morning and take off our clothes and dive in. One time, when we thought we were alone, we jumped in naked and heard a yell off to our right—some guy was standing there fishing and we'd just soaked him. He shouted, "Don't you guys ever sleep?"

Vic also liked the whole lumber industry. We had both worked on every part of the mill, from shovelling sawdust and bark to working on the green chain to driving a forklift and lumber carrier to dogging logs. Machines do this now, but in the early days you'd have three guys on the carriage, two on either end of the log holding a cant hook, which is just a wooden handle with a hook called a dog. We'd dog the log and hold it in place while the head saw removed the sides. Then we'd turn the log, and the saw would take off the rest of the bark. The job didn't call for heavy lifting, but we had to keep our balance. Those initial jobs led both of us to a clearer understanding of how to run a sawmill.

After Vic left B.C. Manufacturing, he'd started working in Nanaimo as a sawyer for MacMillan Bloedel. One day when he was visiting from Nanaimo, he came by to see me. I talked to him about Harold's mill and he listened. I must have sounded pretty certain that we could do it because he kept nodding. We knew our strengths would work in our favour, and the more I talked with him, the more I felt it was the right way to go. The opportunity was in front of us, and we were healthy guys not afraid of work, so why not make a sound investment in ourselves that could pay off in a few years. Vic had a lot of confidence in me and wanted to give it a go, so he said, "Let's do it, Chick."

We decided on the terms. He'd put in 40 percent and I'd put in 60 percent. I didn't want a fifty-fifty split in the business because I wanted to make sure I had the say. Plus, it was my idea. Vic was fine with that. His mother-in-law helped him out with his 40 percent. I went to the Bank of Montreal for a loan and had my dad co-sign. Dad was still working in the mill in Haney, the same mill where I used to pick him up on my motorcycle. He worked there a long time until he was in his sixties.

When the ink had dried on the deal, we had put down $15,000 cash. Interest on the balance of $45,000 was 6 percent per annum. We agreed on a clause that said, "In the event of severe weather conditions, which would curtail production, a monthly sum of $350 would be paid on balance." The agreements were signed by mid-November. A new chapter of my life started in the fall of 1963 when

In 1963, Vic Rempel and I bought the first of S & R Sawmills, A mill.

I bought my first mill with Vic. Our last names were Stewart and Rempel, so we called it S & R Sawmills.

We hired a tally lady to help us keep track of the cuts, which saved us a lot of time. She lived on Barnston Island across the river from the mill, and if the weather was fair to good, she'd row her boat across to come to work. If the weather was bad, she'd take the ferry and get a ride in with one of the guys working at the mill. She'd sit in a cubbyhole by the green chain and would count and write down what was being cut; for instance, she'd mark on her tally ten feet of two-by-fours, and at the end of the shift she'd be able to tell us how much lumber we'd cut that day. We figured out the board footage from her tallies. She was also in charge of answering our one phone while she sat in her cubbyhole.

We were lucky with our employees. Emil was willing to stay on working for us, and the farmers were happy to get steady work, but we had to make improvements before we could get the mill going at full capacity. We didn't have any electricity, so when it got dark in the winter around 4:00 p.m. we couldn't run the mill. The short days were a big disadvantage because we only had one shift working, so profits would stay low. We would have to take out the steam engine and borrow some money to put in electricity. It was a shame to switch because that engine did run the mill well, but we would always need a ticketed steam engineer to run it, and we felt the steam was holding us back. We wanted to keep up with the times, so we shut down the mill and started the upgrade. After about three months, everything changed: steam was out, electricity was in and we were running two shifts on yellow cedar.

We decided to keep spending money to put in good machinery until we perfected our mill. We always kept our eyes open for equipment. We had a circular saw, but because of the design, the saw could cause a defect in the wood: one blade was on top of the other and the teeth could leave a mark in the wood where the two blades met, so we wanted to change from the circular to a band. Luckily, it didn't take long to find a good used band saw. Everyone in the mill business gets a band saw after a while.

We were fortunate to have Emil; he could be trusted to run the mill. He was a hard worker and the heart of the mill. He was also an excellent woodworker, and when he retired in 1980, his daughter, a musician, had got hold of some blueprints for a harp and asked him if he'd build her one. He attached thirty-six strings to a soundboard of yellow cedar and made the curved body from maple, and in two months she had her harp. He enjoyed building that harp so much he built another thirty over the next two years and was making some money from his hobby. We knew he was a musician because he used to play the Hawaiian guitar. He was a talented guy and a terrific asset to our business.

When we started operations at A mill, Vic and I had to put in a few extra hours. We were working six days a week. During regular hours, Vic was the sawyer and I was the yard foreman doing the

Loading lumber at A mill on a Saturday.

tallying, but on Saturdays, when the other workers were off, we'd drive over to the mill and make shiplap on our planer for subfloors. We'd plane up enough lumber to fill a boxcar. The railway tracks were beside the mill, so we'd move the train car into a better position on the rail line and start filling it with lumber. It usually took us about eight hours to get the job done.

Even though we were working hard, we were also busy hunting for a gang saw. The saw is a series of saw blades, toothed on one edge, that cut only on the downward stroke. When the log comes into the mill, the head saw cuts off the side slabs and squares it. This newly shaped timber was called a cant. The cant could be four or six inches thick and ten or twenty inches wide. With a gang saw we could run cants through cedar and, depending on the size of the cant, we could put spacers in the saw to space the cuts anywhere between one and one-and-a-quarter-inch slices, so we might get six or seven boards at the same time, like slicing a loaf of bread. This saw made cuts faster and more precise. The blades had to be changed at least once in a shift, so we'd do that during the lunch or dinner break. We wouldn't use the gang saw all day, just for certain cuts for specific orders, like when the manufacturer wanted to take the boards to his kiln and finish the job there, but the gang saw would give us more versatility in the cuts.

We asked our contact on Vancouver Island to keep his eyes open for the saw. Carl Reid was a used equipment dealer on the Island, and we'd already bought a couple of lumber carriers from him.

Our next purchase from him was a log splitter. Logs coming from the boom in the water were too big to go to the mill and be cut, so they had to be split down the middle first with a splitter that floated on the water. But those two pieces were still too big, so they were flopped back in the water and split again. MacMillan Bloedel had hired Carl to take apart their Hillcrest mill, and he called to say he had a log splitter for us if we wanted it. The mill was being pieced off, and Carl said MacBlo was looking at the log splitter but hadn't taken it yet, and he thought he could sell it to us. Marilyn and I drove over to Honeymoon Bay on Cowichan Lake to take a look.

The log splitter was in the yard and looked good to me, so Carl and I started negotiating the price. Marilyn had gone into his bungalow to put on some coffee, but she ended up cleaning the place. Carl was a bachelor and he'd left dirty dishes in the sink and a bit of a mess on the kitchen table so, while we settled the deal, she cleaned and made us lunch and fresh coffee. We loaded the splitter with a forklift and put it on a flatbed truck on a float out in the boom and delivered it to our mill. Over lunch, we asked him to let us know if he happened to hear of a used gang saw coming available.

Not long after our trip to the Island, Carl called and said he had found a gang saw. He had been up to Houston near Burns Lake and

LEFT Our first company truck.

RIGHT, TOP S & R's original office.

RIGHT, BOTTOM One of our first Christmas cards was this drawing of A mill and our beehive burner. The inside read, "One of the joys of the Season is the opportunity to put aside the routine and customs of every day business and in real sincerity wish you a Merry Christmas and a Happy New Year. S & R Sawmills Ltd." Vic and I signed every card we gave to our customers and employees.

noticed a used one. They were hard to find, so Marilyn and I drove six hundred miles to Houston in my Chev pickup. We had Carl's directions and drove to the warehouse without any problem. We liked the saw, so we paid around $10,000, a fair and reasonable price for a used one. We loaded it onto the truck and brought that piece back down to our mill.

These initial expenses paid off when business started coming in through word of mouth. Guys in the business were saying, "Chick's doing custom work in his own mill." By then, Vic and I had been in the mill business for twenty years, and guys knew we were hard workers, so they sent work our way.

Our first cheque was $4,000 for a pretty big cut. When we drove over to the Bank of Montreal in New Westminster to deposit it, we kept passing it back and forth, staring at it and grinning. We could hardly believe it. Within six months of opening our doors, we were getting steady cuts.

When I had first approached Harold Morrison about buying the mill, he included an interesting object in the price. A couple of years before Vic and I made him an offer, another guy had shown up to buy the mill. The deal was almost completed when this prospective buyer delivered a large safe to the main office and left it on the back porch. The safe was about four feet high and a couple of feet wide. The man never returned, so the deal collapsed, but Harold was always curious about what was inside.

"Why don't you open it?" I asked.

"Don't know the combination."

"There could be thousands of dollars in there."

"Could be."

"You don't want to take it with you?"

"Nope. You're welcome to it if you want it."

Over the years, we kept the safe tucked away on the back porch, never bothering to try to open it. Customers and employees would see it and ask, "When're you going to open that thing?"

"Not sure."

"What do you think's in there?"

"No idea."

Sure, I was just as curious as the next guy, but the safe had become a topic of lively conversation, and I enjoyed hearing all the imagined possibilities. A few employees even tried to work out the

Left to right: Colleen is three years old, Suzanne is eight, and Wendy is twelve. This photo was taken at our home on Hammond Ave. in Coquitlam.

combination but had no luck, so I left it sitting there for thirty years before finally hiring a locksmith. But that's a story for later.

Our little family had grown to three beautiful daughters—Wendy, Suzanne and Colleen—and I was beginning to believe our ability as parents to raise our girls in financial security would happen if we kept on this new path. Our success would depend on Marilyn and me doing what we had learned as kids: if we worked hard and counted our pennies, we'd all be okay.

After that first cheque of $4,000, our business continued to grow. Barry Tyrer from Trans-Pacific Trading found out about us through word of mouth. Barry's company had been exporting lumber to Japan since the mid-1950s, and he helped bring us our first overseas customers. He came out to see me and got us an order from Japan for four-and-a-quarter by four-and-a-quarter-inch squares of yellow cedar they used for their construction. The Japanese favoured yellow cedar (in the bush it's called cypress). The wood is durable and lasts a long time. They wanted large orders of the four-and-a-quarter-inch squares so, in mill terms, we'd say we cut *heavy* for those.

The two Japanese companies, Emachu and Sanyo Pulp, made a big difference in our production. The Japanese population was growing fast, and the government didn't think their lumber industry could meet the needs of its people. The BC forest industry had been looking for new markets because US markets were drying up, so the partnership between the forest industry in BC and S & R was beneficial for everyone. The orders were increasing fast, so we had to make sure we had two good shifts to cover it all. Soon, both Japanese companies began to fight over who would get the first position on the cuts. In the beginning, when we didn't have as many customers, the two competitors kept a close eye on the tallies. If, for instance, at the end of the year Emachu had got four shifts in a row and Sanyo had got five more, we would make up those five to Emachu, so they had equal time in the mill. We tried to keep both of them happy. Sanyo might ask, "When are you going to get off Emachu's lumber and start on ours?" If this was a Friday, we'd tell them, "It'll be on Monday." We wanted to be fair. Everything was on a handshake basis, no contracts, so they relied on us to be true to our word.

We would give both companies an approximate price. They'd say they wanted a certain number of thousand feet and we'd tell them the cost. We might have a million feet of yellow cedar, and we'd give them a schedule for when we could start cutting. We tried to cut the same amount for each customer. After a big cut, I usually found a bottle of Scotch on my desk.

As our company grew and we got more customers, everyone was on a first-come, first-served basis. Emachu and Sanyo took up a large chunk of cutting time, so when we branched out and bought another mill, we were able to devote the time to their orders in that mill. When we finished their orders, we'd load the trucks that took the wood to the dockyards where the Japanese ships were waiting.

Another company soon arrived at S & R for custom cuts. Sam Hughes and Bill Gillis were in charge of sales for Mill & Timber Products in Surrey. Mill & Timber became one of our big local customers. The company was making bungalow siding and wanted red cedar cut heavy, or as much as we could get from a cant, into what we called gang boards. This was the most widely used wood siding, made by recutting the lumber at an angle to make thicker pieces on one edge. One side was textured and the other side smooth. We started a process at our mill where we cut several planks through one piece of board with our gang saw. For instance, we had the gang saw cut a ten-inch-thick cant of "heavy" one-and-a-quarter-inch red cedar for Mill & Timber. We'd get maybe ten-inch-wide boards up to twenty feet, and the guys at Mill & Timber would take the boards to the dry kiln at their mill to season it and get rid of the moisture, and then they'd run it through their re-saw to get the angle for their one-and-a-quarter-inch by ten-inch bevelled siding. We didn't do the bevelled siding. We just made the blanks (mill lingo for boards).

S & R Sawmills is a custom-cutting operation. We never owned the logs. Customers had their own logs and would specify how they wanted them, and we'd convert them into the lumber products they'd ordered. In those early days, we were cutting eighteen thousand board feet of daily production with eighteen employees.

We had contracted out a few employees to do one of the most important jobs at the sawmill. These guys were the professional saw filers working in the filing room. I never did any filing myself because it's a specialized trade in circular and band saws, and the guys we hired had been trained for years. The sawmill saws are large and expensive, and they need precise care to make sure the whole operation is safe. A dull or broken saw can wreck the lumber, and filing it the wrong way can destroy the blade. When you're running lumber through the mill, the saws have to be used carefully. In the mill, we'd change the saw every four hours; the removal of one and replacement of another took about five minutes. We'd take a band saw off with a crane and lay it on the floor. We'd try to change the blades during break time, but if a saw started to go dull or had a bent tooth, we'd ask the guys to go on their break a little early, so the blade could be fixed and we wouldn't lose production. The workers never wanted to lose time, so they always adapted to the change.

In the beginning, we didn't have enough saw filers of our own, so we contracted out for the workers. Walter Chaston had a group of trained filers working for him, and he would send over as many as we needed to run the filing room. Depending on what was wrong with the saw, it could take about an hour to get the work done. The filer might need to weld a crack in a blade or just sharpen the teeth. Guys can get hurt filing, but they wear gloves, and special machinery loads the saws onto the workbenches to make it safer. Now we train our own filers, but when we were starting out we called on Walter. Filing is one of the highest-paid jobs in the mill, so a young fellow starting out in the file room was learning a good trade. The head filer has four guys working for him, and each mill has a head filer and a head millwright. We soon asked one of Walter's guys to train our own employees, which he did, so we don't contract out the filing work anymore.

One incident that happened during this time involved one of our saws. I was home asleep, and around 2:00 in the morning the phone rang. Someone from the mill was calling to say a saw had come off the big log splitter. The guy on the other end of the phone asked, "Where can we get another saw in a hurry?" I told him to leave it with me. I knew the spare was at the shop where we'd sent it to get sharpened, so I phoned that guy to see how soon we could get a replacement. He said he'd have one ready for the morning.

As I sat up in bed I imagined what had happened. The log splitter was on a floating platform in the water, and the blade that looked like an eight-foot crosscut had fallen into the river. We had to get some logs bucked (cut) for the morning, so the men were splitting the logs, but when they changed the saw, it fell into the water. I knew they would be getting another splitter bar in the morning so they could keep on schedule and continue cutting, but I couldn't get back to sleep. That saw was in perfectly good shape. Plus, it was my saw and I wanted it out of the water, so I got dressed and drove over to the mill.

When I arrived, the spare was already out there and the men were working. I went to the edge of the water and asked them, "Where did it land?" They pointed to the water right below the float. My plan was to take advantage of the diving course I had recently finished and to wear my diving equipment to retrieve the saw. I put on my tank and stepped over to the edge. This was my first experience without my instructor, but I could see the water wasn't deep. While on the course, we were taught how to breathe properly underwater. Unfortunately, during the training sessions, I'd swallow so much water that I'd be uncomfortable for some time as it sloshed around in my stomach. I wanted to avoid that happening, so I made sure my breathing apparatus was on properly.

The water was only about ten or twelve feet deep, but bitterly cold and dark. We attached a rope to the splitter deck for one of the guys to lower me into the water. When I got to the bottom, I couldn't see a thing, but when I looked up I could see the light above, so I knew I was close to the surface and would be okay. My breathing was fine, so I started searching for the saw. I was as good as blind down there, so I spent a few minutes feeling around with my hands until I

TOP A tug towing booms from the BC coastal forest up the Fraser River to S & R storage grounds.

BOTTOM Bundles of logs at S & R storage grounds. DETLEF KLAHM.

found the saw. I was busy tying the rope to the hole in one end of the saw and, of course, couldn't hear anything above me.

The head boom man saw the bubbles coming up to the surface and thought I was losing air. He called to another guy for help, suggesting one of them jump in and rescue me. Before anyone else came in the water, I surfaced, using the rope to guide me. The men were glad to see me, and one helped me climb onto the deck while

I'm getting the log ready for the log splitter.

the other slowly pulled up the saw. The saw was pretty heavy and took two guys to lift it, but we got it out of the river and were able to use it again. This took about half an hour, and I was shivering like crazy. They put the heat on in the boom shack, and I sat in there getting warm by the woodstove. Took a while to get rid of the chill, but my stomach was clear of water and I had my splitter saw back.

Usually, an owner doesn't have to dive in rivers to collect his saws, but the running of a sawmill *can* be a complicated business. The whole process started with the red and yellow cedar booms floating down the river to our mill.

The logs could come in anywhere from one to eight sections of flat booms or bundle booms. The bundles had a strap around eight or ten logs; we'd open the strap and the logs would spill out. One section could contain twenty thousand feet of timber, forty feet by forty feet. The sections could be a mix of sizes—some peewee logs and some big logs.

The way we cut the log depends on the customer. If we had a forty-foot log, the customer would want three pieces each of thirteen feet (three-by-thirteens), so we'd get a big cross-cut saw or chain saw and cut it in about ten seconds. When we cut a log into these shorter lengths, it's called bucking the log. We had to gain experience to figure out how to cost out a custom cut. If a boom was five sections, we'd know approximately how much we could cut in a shift, but there could be flaws in the logs and different lengths.

We still use different methods to get the logs from the water to the mill. We might have side lifts or a jack ladder to carry the logs into the mill to the head saw. A couple of guys might be on the boom guiding the logs, or one guy could be in a small boom boat while another one guided the logs with a pike pole (a long pole with a hook) onto a log lift called a jack ladder. This ladder is a rotating chain conveyor that grabs a single log and carries it vertically to the upper levels of the mill.

The log slip doesn't have the chain conveyor. On a side lift, the log gets caught in little buckets like metal feet to keep it from sliding back down, then slides up horizontally until it spills onto the deck before it gets rolled for loading onto the carriage. The side lift continues in a circular motion, catching the next log. A guy at the top cuts off the bark with a barker before the log is rolled and dropped onto the carriage that takes it into the mill. The boom boat plays a role too.

If there's a bundle of low-grade logs from the boom, the boom boat pushes the bundle from the water onto a steel plate that sends it up the side lift to the crane up top, where it cuts the binding and spills the logs onto the deck. Then a worker leads the logs over to the chipper, where the junk logs become useful chips. The chips travel down a long chute and spill onto a scow in the water.

We have always calculated the cost of the cut of the lumber in two ways. Our customers paid a shift rate or paid by the thousand feet, whichever produced the greater amount. The better the logs, the more we could cut, and the more money we could earn. When the booms were opened at the water, we checked for lengths and defects. Clears were the highest grade with no knots or defects, so they were worth more, and they didn't slow down the process. When the logs were poor—a bunch of junkers—we went by the shift rate because the junk logs took longer to cut. The minimum rate for an eight-hour shift was around $20,000, but $25,000 per shift was really good. Clears could be four and a quarter, and wider lengths heavy to thirteen feet. Sometimes we'd take the forty-foot log up to the deck to cut it. We could buck a forty-foot log into two twenties or three thirteens. If it was damaged the flaws would be close to the butt—the large end of the log—and the customer would say, "We'd like heavy to thirteens and twenties," so with the defect, we might get two thirteens and one ten, but we'd get good lengths of thirteen feet and twenty feet, no matter. We could get 120,000 feet or 40,000 feet on a shift, depending on the grade and size. My tallying exams prepared me for all the calculations to get the numbers right: we never wanted to be short-cutting ourselves.

When we bought the mill, we had promised to pay Harold $1.50 for every thousand feet of cut lumber, which we thought might take a couple of years, but when we showed the bank manager our reports of the increased cutting, he told his assistant to lend us the money to pay off our debt to Harold sooner than planned.

In 1966, Bob Greenwell's mill in Pitt Meadows, called Greenwell Blackstock, was looking for a buyer. The mill was small, cutting red cedar, so we talked to Bob about buying it. We made a deal to purchase the mill and its two acres of land on riverfront property for $30,000. Nine days after the sale was completed, the mill suddenly burned down. We never did find out how the fire started, but luckily we had insured the business for $30,000. In those days, the area lacked proper water services, so fires were a common hazard in the mills. The insurance company investigated and concluded the fire was an accident and offered us $30,000 in replacement costs. We asked the insurance adjuster to pay Bob Greenwell that sum, which would close our deal with him.

The Greenwell Blackstock mill had a crew of twenty men who were now out of work. We hired them for A mill and soon had three shifts running twenty-four hours a day, five days a week. The only disadvantage was that the new mill property was on the other side of the Fraser River, across from S & R Sawmills, so some of the men had to drive pretty far to and from work each day, but some of them carpooled. They didn't mind; they had a job, and they helped us get the third shift going.

One of the workers from Greenwell was Ron Drew. I had first met him in Tahsis, and he was the guy we always dealt with at Greenwell, so I knew he would make a good foreman. We offered him the job and hired him to run the afternoon shift.

In 1970, Tom Pallan of Pallan Timber decided to move his forestry business from Port Kells to Campbell River on Vancouver Island. He had a mill and fifteen acres downriver from our A mill, but his mill had burned to the ground. He put it up for sale and we bought it to get the property along the river. We decided to build our own mill on the site, but we needed to spend some time cleaning it up before we built the footings.

I guess you could call it a twist of fate when, in April 1973, we heard about the next big sale. After fifty years in the business, Nalos Lumber, the company that had fired me, had gone under, and the sawmill was going up for auction. Mr. Nalos had been the only owner in the area who cut the big timber exclusively. His mill had one of the largest head saws on the coast that could make boards out of a log six feet in diameter and split larger logs into quarters. B.C. Forest Products owned the Nalos Mill at the time and was losing money, so now the contents were on the auction block. Vic and I were mostly interested in that giant band saw and some other components, so we went over to the north shore of False Creek at the foot of Smithe Street to see if we could make a couple of bids. Sure enough, we found some items and started bidding on the equipment. We bid on the band mill, the carriage and edger, the transfers (the chain conveyors), and the head saw, and got them all. We were worried, though, that the saw might get damaged in the demolition. Then we saw that Nalos's two-storey building, the heart of the mill, was also for sale.

We put our heads together and figured this might be the perfect mill to put on the twelve-foot-high footings we were building on our fifteen acres, so we bid on the Nalos main building and got that as well. No one thought we could take the whole building out, but we discussed it with Ron Fielding of Industrial Timber Mills—a first-class outfit—who said we could move it all in one piece with the machinery intact. His plan was to put the building on a flat scow and ship it downriver on a barge. He and his crew cleaned out the bottom of the band mill, dismantled the top and bottom wheels of the band saw and laid them flat on the floor of the mill before loading it on the scow.

The barge began its journey out of False Creek and toward Port Kells. An old railway bridge had to open its span to let the mill float through. The newspapers covered its passage and called it "The Great Escape."

That day, I stood on the shore and watched what would soon become S & R's B mill as it travelled toward its new location. If Mr. Nalos had never fired me, maybe I'd have been witnessing this move as one of the workmen now laid off at that mill. The thought did cross my mind that it was a good thing he'd fired me.

I got in my truck and met the scow as it arrived in Port Kells three hours later. Fred Fuchs, the head millwright, had already been preparing the foundation. (The five Fuchs boys followed their father in the mill and spent a number of years at S & R; they were all good workers.) We started major renovations that included adding a timber deck out one side so we could cut large timbers up to forty feet long, but the construction wouldn't be finished for about four months. B mill became our largest mill and specialized in larger sizes and cants, cutting mainly spruce logs, but we could also cut hemlock, cypress and fir.

In the meantime, I continued to keep my ears open to any new prospects, and I soon heard Rod McKerlich was looking for a partner in one of his mills. This sounded pretty good, so he and I had a talk. We had no idea our first discussion would lead to a lifelong friendship or that the two of us would inadvertently blow up part of the Fraser River.

6

EAGER TO EXPAND

N 1970, two acres of property came up for sale along our mill's
road on 98A Avenue near A mill. Rod McKerlich was a good
sawmill man; his trade was as a filer, and he wanted to go into
partnership with me. He was born in Vancouver and had served
in the Canadian Navy on ships going from Newfoundland to Ireland, so he welcomed adventure. He had made good money running
a sawmill in Oregon for a few years, but missed his homeland, so he
packed up a few truckloads and crossed the border with his mill. "A
bad idea!" he told me. "Too much work."

On Rod's first trip, he loaded the truck with all the saws. The
truck was weighed down so much that he came through a section
of Seattle going about five miles per hour with a long line of cars
behind him, but he was determined to make it to Blaine, Washington, even though the stress was getting to him. He slept in a motel
in Blaine and dreamed his truck rolled over him and he had to
climb out the driver's window before the truck caught fire. "A terrible nightmare," he said. That didn't stop him; he made three more
trips with the remainder of his mill equipment before he settled
in Port Kells.

At first, Rod and I built a stud mill and called our company Stu-mac—a combination of our last names, but this site would later become S & R's C mill. In those early days, our stud mill was a basic building, cutting mostly two-by-fours for framing and moving along the green chain, where we graded the wood.

We never got permits for anything we built, and one day the city manager for Surrey showed up and asked, "What's that building?"

"We just put that up."

"Did you get a permit?"

Rod turned to me and said, "I thought you were going to do that."

I said, "I thought you were."

The city guy shook his head. They were glad to see us come into the area because we created employment, but they wanted us to fol-low the bylaws. He let it slide that day, but he had a special request. "I'm not going to ask that you get a permit for this, but you guys have to stop peeing in the bush. Build an outhouse!"

In 1970, anti-pollution regulations brought in changes. If you travelled to small-town sawmills, the tallest structure would be the burner, seen for miles. Unfortunately, even though these burners meant the lumber industry was doing well, they spewed too much smoke when mills burned their scraps. The heavy output of smoke went against the new anti-pollution standards, so those old burners had to go.

Harold Morrison had a beehive burner when we bought his mill in 1963, but we replaced it soon after. One morning, my brother Herb noticed an RCMP constable standing in front of the burner with a box containing a uniform and boots. At one of the access openings at the base, he threw in a pair of black trousers.

"You planning to burn all that?" Herb asked. "It looks new."

"Have to." The constable threw the red jacket into the fire.

Our beehive burner before it was phased out in 1972. Bhana, A mill's watchman, and his dog Laddy. The small shed under the conveyor was our blacksmith shop.

"Doesn't anybody want these clothes? Seems a shame to burn them."

"Somebody might use them to impersonate an officer. We have to destroy them."

Herb stared at the sturdy, black boots lying in the box. He picked up one and admired it. "This too? You gonna burn these perfectly good boots?"

The constable took the boot and threw it into the burner.

Just before we got rid of our beehive burner, I had come into work about 7:00 a.m. I saw that coals from the burned timber were simmering fine, but they'd scorched out a saddle shape. In the open air, if the wind had whipped in, the coals would have flared up and burned down the whole mill, so we were happy to say goodbye to the beehive burner. When the government ordered the mills to start phasing them out, Rod and I decided to build our stud mill with a modern design.

Wood buildings and sawdust can be a dangerous combination. The steel construction of our new mill was meant to reduce the fire hazard, and the giant conveyer under the main building would carry the sawdust out to a bunker about a hundred feet away. We'd drive a truck under the bunker and open a trap door that would spill all the sawdust into the back of the truck.

At first, we took the truckloads of sawdust to Bob Poulevard, who bought and delivered it to chicken farmers. Bob was one of the guys in the hospital years before when I had my accident, so I'd known him quite a while. Bob's business helped keep the bunker empty. Soon we got so busy that we relied on him to truck the sawdust out from the bunker. We'd phone him to come, and he'd get three loads a day out of there. It was busier when we were running two shifts, sixteen hours a day, but we also needed it hauled out at night. Bob hired a truck driver to do it at night. This system worked well until, eventually, we used conveyers to run the sawdust to the piles in the yard.

Fire was a real threat, so smoking in the mill was grounds for dismissal. Our employees knew to watch for sparks or friction, hot objects or a direct flame, anything that could ignite sawdust, wood shavings and chips. We always had to be careful with welding equipment, cutting torches, faulty wiring or a blown fuse. We figured we were protecting the men and the mill with the conveyor getting rid of the scraps. Fortunately, we figured right, and this mill never caught fire.

Rod and I used to fly to parts of BC to buy used equipment. We never bought new because it was too expensive; everything was used. One time we flew into Edmonton in the middle of winter to look at some equipment, and we made it back in one day. The whole place was freezing cold. We landed and rented a car to drive to a lake to meet the salesman. The guy was standing by the lake waiting for us when a storm blew in, which made it even colder. The choppy waves froze in mid-air. We'd never seen anything like it—must have been −40°F.

On our drive back to the airport, we got stuck in the snow near a weigh station. A couple of guys helped us push the car and told us that a few months before, a farmer had been picked up for having an overweight vehicle. He was so angry, he went home and filled a forty-gallon drum with warm water and poured it into the scales' innards. Froze the works for months.

We were planning to stay the night in Edmonton, but luck smiled on us when we returned the car to the airport. An Air Canada flight had just come in from Europe and was on its way to Vancouver. We got the word that they had two empty seats, so we hopped on to get home and get warm.

We flew over to Vancouver Island and inland BC a few times to buy used equipment. On the Island, our pilot usually landed in Cowichan Lake, but this time he noticed a small millpond was empty of logs, so he landed the floatplane in there. The guy at the mill said no one had ever done that before. We soon found out why. The problem came when we tried to fly out, and we didn't have enough distance to gain proper altitude. Rod and I kept staring ahead as the plane approached the front door of a farmhouse. The pilot was struggling to climb, and we figured we'd be crashing through that door any minute. We braced ourselves for the impact, but the pilot managed to lift the nose and get the plane high enough that we made it over the roof in time.

I turned to the pilot and asked, "What would you have done if we didn't get high enough?"

My 1971 Chev pickup parked beside high-grade Douglas fir logs in the mill yard. DAN KETTNER.

"Cut the engine." He figured crash landing in their front yard was the better option. After that incident we avoided landing in the millpond.

A couple of months later, the pilot landed our plane in a farmer's field at 100 Mile House. Rod and I were there to buy a German gang saw. When we headed home, part of the plane's dashboard fell in my lap. The pilot told us not to worry: "I'll fix it up when we land." Rod and I were both thinking, "*If* we land."

These minor mishaps inspired me to go for my pilot's licence. I bought a Beaver floatplane that I kept in Fort Langley and used it when we needed to look at logs in the water. I'd hired another pilot who also worked in the office, and I'd say, "Norm, we're going to see some logs in the Queen Charlottes," and he'd ready the plane and off we'd go. After a while, I bought a Cessna two-seater and started taking flying lessons on the weekends. I was getting the hang of it, and one morning I told Marilyn to come out to our backyard to see me fly over the house. I buzzed overhead and saw Marilyn waving a white sheet, so I tipped my wings toward her as I flew by.

My final test was coming up, and the instructor took me through the route I'd be flying the next day. He told me I would need to fly solo in order to get my ticket. This unnerved me a little, but I focused on the practice run to make sure I knew what to do.

We fired up the plane and took off from Langley airport to begin our "touch and go landings." The weather was clear and fine, a beautiful sunny day, so the journey to Abbotsford, Chilliwack and then Hope went well: so well, in fact, that the pilot was relaxed enough to sing Irish songs most of the way. He had a good voice as he belted out "When Irish Eyes Are Smiling" and "Danny Boy." The singing kept me calm. As I completed the circuit and taxied onto the Langley runway, he said, "Barring any unforeseen circumstances, you'll do fine."

As I set off the next day for my solo test, my instructor radioed to the three airports and gave them my flight details. I left Langley and headed toward Abbotsford. All was good when I touched down at the airport and took off again without coming to a full stop. Same thing for Chilliwack. I was getting so relaxed I was about to sing one of those Irish songs, but then the problem began.

I was flying through the mountains from Chilliwack to Hope when the plane suddenly hit storm winds. The combination of downdrafts and valley winds forced the plane to drop about fifty or sixty feet. The winds buffeted the wings hard. The conditions were horrible, but I gripped the controls and tried to think: Should I turn around or keep flying through it? I chose to stay on the flight path; circling back might be more dangerous. As I approached Hope's runway, I dipped lower than normal and flew beside the railroad tracks. I came alongside a train going the other way and waved at the shocked engineer. I figured he might be the last guy who saw me alive.

The plane had no radio communication, so I touched down in Hope and headed back to Chilliwack. I had doubts about this plan, but went ahead anyway. I wanted to get home. Fortunately, by the time I got to Chilliwack the weather had cleared.

When I exited the plane in Langley, I had made a decision: I wouldn't be doing any solo flights. I had passed the test but never took the written section of the exam, so I didn't get my licence. It wasn't worth it. Sam Hughes, who brought us Mill & Timber Products back at the beginning of S & R, had crashed and died while flying his floatplane from the Gulf Islands to Vancouver. The risk was too great. Plus, Marilyn didn't want to fly with me, and she was also afraid I'd crash, so I thought it a good idea to hang up my wings. Over the years, I would fly with our pilot and if he had to do something he'd say, "Chick, take the controls." That was enough for me. I like adventure as much as the next guy, but I didn't want to kill myself.

Rod McKerlich and I kept our adventures at Stumac on the ground. One year, the Fraser River froze over and we couldn't move the logs in the boom, so we made a plan.

We hopped in my truck, drove over to the co-op in Langley and asked the guy if we could buy some dynamite. He took us out in the snow to a locked shed behind the store and gave us a full box. We went back to the mill and got a drill. Rod drilled the holes in the ice around the logs, and I placed the sticks inside. We lit the fuses and a few seconds later the ice exploded with a *BOOM!* A bunch of salmon flew over our heads and flapped onto the ice. We managed to catch a couple and were able to free the logs. To quote Rod: "That was the craziest thing we ever did."

The frozen Fraser River out front of S & R.

The surrounding area in those early years looked different from today. Instead of sawdust piles and stacks of lumber, there were three houses on our main road on 98A Avenue, and a lime (the mineral) plant was doing business nearby. They brought in raw material for the plant from Texada Island up the coast. A fish-smoking plant on 192nd Street just north of 96 Avenue kept busy smoking local and imported fish.

Vic, Rod and I used to go for coffee at a couple of cafés on the other side of the tracks near the entrance to our mills. We met Howard Thompson there. Howard lived across from our mill and was our watchman for years. He and some old-timers, guys over eighty, used to drink their coffee and tell us about the early days as far back as they could remember. We heard these stories so many times, we had them practically memorized.

In 1859, a guy named Donahue built the first sawmill in New Westminster. Around 1900, two guys owned the first sawmill in Port Kells, and the sawyer was a guy named Parsons who lived on 192nd Street at the corner of 94 Avenue. The logs came from hand logging along the river and on city lots. Those guys could make $1,000 for clearing an acre, a lot of money back then. They used the wood for homes, barns and a couple of hotels.

"Some of the largest timber on the coast was in Surrey."

"In 1893, a four-foot-square timber, 105 feet long, got shipped to the Chicago World's Fair. We had a fir tree measured eleven feet round and thirty-six feet from the butt."

We heard from these old-timers that a rich guy put Port Kells in the news. Before World War I, around 1910, Baron von Mackensen built a big house called the Castle on the southeast corner of 96B and 192nd Street.

"He'd invite the whole community to his place for a Christmas party. Everybody went—men, women, kids. We all got gifts. His parties were the highlight of the season. Trouble started at the beginning of the war. The Baron raised a German flag on his roof, and the councilmen told him to take it down. Found out later he was a spy! The papers got hold of the story and that ended the parties."

During World War I and into the 1920s, a couple more sawmills were built on the same patch of land as S & R occupies today.

"Mills came and went. Some were shingle mills, but most of them burned down or went broke."

"The New Westminster Southern Railway was here, don't forget. It came in 1891. Made it easy to harvest timber in Cloverdale and along the Fraser. Oxen dragged the logs to the river and then again up to the tracks. BC Electric Railway came in around 1910 and the northwest uplands of Surrey got opened up. Most of the mills were next to the railways. Lumber got sent via BCER to Chilliwack and the Canadian Northern Railway to the east, and across the West-minster Bridge to the docks on the Fraser, or to the Canadian Pacific Railway."

"We had a timber baron running things for a while. James Good-fellow Robson owned The Timberland mill. Robson was a lumber magnate from New Westminster. His first job was as a millhand at a sawmill in Haney, and he became partners with the owner on a shingle mill. At twenty-three, he was broke and out of work. He'd gone bankrupt in the lumber business after two years. He got lucky, though. A new mill started and needed a manager."

"Timberland could cut forty thousand feet of lumber in ten hours. They had around seventy-five guys working for them."

J.G. Robson was well known in our community. He turned his lumber business into a success for the next fifty years. In the mid-1950s, he donated $100,000 to the Royal Columbian Hospital, and made several big donations to his church and to the University of British Columbia. Vic told me Robson gave the New Westminster YMCA $200,000.

Pretty much all the timber was gone from this area by the 1930s.

A few years after I bought the mill from Harold in 1963, our mill neighbours included two cedar mills east of 192nd Street, three lumber mills west of 192nd and a mill that chipped hog fuel. Gradu-ally, more industry flowed into the neighbourhood, and the look of the place started to change. By 1977, I was ready for change too, but not this kind: once again, fire threatened to destroy a business, but this time I found myself inside the burning building.

In 1977, I told Rod I wanted to expand the stud mill and make it a proper sawmill. At the time, S & R was the smallest mill in Port Kells, while Fraser River Sawmills on the northern shore of the Fra-ser River and Timberland Lumber Company in Surrey were big mills, and I was thinking about growth for the company. Rod was ready

Rod McKerlich and I are celebrating our fifty-second Christmas lunch in a row. He's the tall, good-looking guy on the left.

Lumber off the green chain and timber deck has been sorted for grades and sizes before packaging.

to move on by then and asked that S & R buy him out. We'd been fifty-fifty partners, but when Vic came on board, he agreed to our usual sixty-forty.

Rod has remained a lifelong friend and we still meet for lunch every Christmas. In 2016, he turned ninety-three.

In the spring of 1977, construction on C mill began. The building was a Quonset hut with the semicircular cross section. The design came from World War II shelters: easy to put together half shells with height in the middle for the head rig—the saw that makes the first cuts in a log.

By the spring of 1978, C mill was completed. We hired eighty-five employees to work on orders for the Japanese market, cutting the same sizes on the yellow cedar. Now both A and C mills were running three shifts, five days a week. In the meantime, we were still setting up B mill for custom cutting with fir and hemlock logs.

A few months later, in July of '78, I got a phone call at 2:00 in the morning from one of my workers saying the new mill was on fire. Marilyn and I jumped in the pickup and sped over to the mill. When we got there, the flames were so bright against the black night that I figured the building was a goner. The fire truck was there, but the firemen were just standing around, watching it burn. I grabbed my flashlight and ran, shouting, "Why aren't you guys turning on your hoses?"

The head guy said, "The fire's mainly at the back, but it's too dark. We can't see how to get in."

I quickly realized that because we had no I-beams holding up the roof, it had collapsed in the middle from the heat, and it might not be as bad as it looked. This gave me hope that the mill could still be saved. "Follow me," I said. "I know the way."

I led them around the back by the river. The fire was on the main floor above the ground floor, so I guided them up the back steps, shining my flashlight on each step. The firemen were right behind me with their hoses. They were hesitant to climb to the mill and fall through the floor, so I kept encouraging them to stay close to me as we went along the gang walkway. When we got closer, we could see our way inside. We were walking slowly to make sure the boards didn't drop out from under us. If the floor gave way, I'd be the first to go.

We weren't sure where the fire had started, but we could feel the heat and see a mess of flames. The smell of smoke was unpleasant, but I didn't have time to be afraid because I wanted to get those hoses turned on as soon as possible. As we got near the flames, I searched the area with my flashlight and saw that the edger had caught fire. The machine was burning, as was some oil on the floor and the surrounding wood. Within seconds, the firemen got the water running and told me to get out. I was glad to oblige; my throat was burning.

The Surrey firemen fought the blaze for two hours. The fire damaged a cutting room, a plywood room, and a maintenance area, nothing that couldn't be fixed, but that meant the mill had to be temporarily shut down.

The bigger problem was how to keep the employees working. We had insurance on the sawmill for loss of production, and I'd also taken out insurance on the employees' wages so they would continue to receive their pay. But they didn't want to just sit around; they wanted to keep working. A few months earlier, I'd hired Ray Chretien to be foreman on C mill, so Vic, Ray and I organized a meeting for the workers at the Port Kells Community Hall. We figured out a plan to work double shifts on A mill, which meant running on Saturday: we'd employ half the crew for three shifts on Monday, Tuesday and Wednesday at A mill, and the C mill crew would come over to A mill on Thursday, Friday and Saturday for three shifts. They'd work three days but get paid for five. This way, the men kept working and the insurance company didn't have to pay 100 percent for five days lost but, instead, paid for just two days' lost pay.

The rebuilding of the burned section of C mill finished in early '79. Ray and the day shift foreman, Claud Muench, had put their heads together with management and suggested we change the design. We decided to build a square-shaped, multi-storey building for two purposes. First, the increased footage for a bigger room meant we could put in two head rigs to add production. This was a good idea because if one broke down, the other would keep working. Second, we would put the filing room upstairs so saws could be lowered to the head rig in about seven minutes, during a coffee break, instead of the more complicated method of moving them around the band mill sideways with hoists, which took twice as long. Also, with two head rigs in the mill, we could use one to cut larger logs and one to cut smaller logs.

With the improvements and the expansion, we were keeping our Japanese customers happy. The Japanese preferred our yellow cedar because it was genetically similar to their indigenous Hinoki cypress, a small tree that was disappearing rapidly. Funny thing was that for years, loggers viewed yellow cedar as weed trees or log scrubs because nobody wanted it. Everyone wanted Douglas fir or red cedar except for the Japanese. They value certain kinds of wood for exposed uses in homes and temples. Clear, close-grained, light-coloured, vertical-grained lumber is a valuable commodity for posts, ceilings and screen and window frames. This demand for yellow cedar kept A and C mills busy.

In January of 1969, a couple of years after Barry Tyrer's company Trans-Pacific Trading had brought us the Japanese customers, he suggested I take a trip to Japan. I knew no Japanese, but Barry assured me he would be there at the same time, so I agreed to make the trip. Things didn't turn out quite the way I'd hoped.

7

JAPAN JOINS
THE TEAM

ARRY AND I had arranged to be in Tokyo at the same time so he could introduce me to our customers. My main anxiety while I was on board the plane was my lack of any knowledge of the language, but I assumed Barry would introduce me to Japanese clients who could speak English, and together we would be fine; nonetheless, I tried to learn a few words on my own.

As the plane taxied into the Tokyo airport, the flight attendant kept repeating over the loudspeaker to the passengers, "Check the notice board." I thought I'd better look to see if there was anything for me and found an envelope with my name on it pinned to a bulletin board. Inside was a note from Barry: "Chick, I have become very busy tonight having to attend a meeting with the directors of Sanyo Pulp. After clearing customs, you will see three men from Sanyo who will be holding a large sign advertising your name. They will take you to your hotel. I will meet you there for breakfast at 8:00 a.m. and review your week's schedule."

I just about had heart failure because I had no idea where I was going. Also, along with my suitcase, I had a bag with three bottles of Scotch that I was worried might not make it through customs. Barry had advised me to buy it because our Japanese customers liked whiskey. As I approached the counter, the customs guy barely looked at me or at my bag as he stamped my passport and waved me through. I kept going, following the hundreds of people ahead of me.

Finally, I walked into a big area full of people, growing even more nervous about my next step. I hoped that at least one of the Sanyo men could speak English. I made my way to the arrival section and immediately saw three men in the distance holding a giant poster with my name on it high above their heads. They lowered their arms and then raised them and shifted from one foot to the other, so I gathered they had been standing like that for a while. I made my way through the crowd and stood in front of them. I pointed to myself and at the poster and smiled and nodded.

They seemed genuinely happy and relieved to see me. While one of them quickly crumpled up the poster and threw it in the garbage, the other two gestured for me to follow. Not one English word was spoken. We got my suitcase and then stepped into a car that was parked close to the exit. One drove, another was in the front passenger seat, and I was in the back with the third man. I wondered how we would communicate, but my concerns weren't necessary: within one minute of leaving the airport, the two passengers had fallen asleep, so it was a quiet ride to the hotel.

The next morning, I found Barry in the hotel restaurant. I felt reassured until he explained that he had too much work to do and could not accompany me to see the customers. He gave me my ten-day itinerary, which outlined where I would be going and who was going to see me. He then told me he would meet me back at the restaurant in exactly ten days to hear how I got on.

I studied my itinerary and hoped for the best. I had learned early on from my days selling *The Saturday Evening Post* in downtown Winnipeg to always look forward and believe that I can do it. I chose to take this as an opportunity to test myself in uncharted territory.

Over the next ten days, I travelled quite a ways to meet my customers, mostly by train and bus, rarely by taxi. One afternoon, I was riding on a bus with a group of school children who quickly gathered round me to practise their English. The children, like my customers, were polite and welcoming, so my weakness with their language never became an issue: someone always knew enough English for us to communicate. I started to relax and take in the beauty of the country and its traditions.

Their tradition of politeness came clear when I went pheasant hunting with a Japanese fellow. He heard I liked hunting, so he lined up an outing for us. A couple of men were there, and one man had his hunting dogs. I was holding my loaded shotgun, walking with the others and hoping to see a pheasant, when suddenly the dogs stopped in their tracks and started barking. A flock of pheasants flew up to the sky, and we fired our guns, *bang, bang, bang*. One of the Japanese guys yelled out what roughly translated to, "I got him! I got him!" The man with the hunting dogs poked this man on his shoulder and said in English, "No, Chick got him." I'm not sure who actually shot the bird, but they kept it for me and had it stuffed and mounted before sending it to me a few weeks later. In fact, I still have that pheasant hanging on my wall.

On that first trip, I learned a few things from my customers—particularly, what cuts they wanted in the wood. They were remanufacturing the lumber that we cut for them. Before this trip, we always cut the forty-foot logs at thirteen, thirteen, and fourteen. We knew they loved the thirteen-foot, but with a defective butt we'd trim to twelve. Now I saw a guy cutting down the thirteen to twelve and a half. We didn't know they were using twelve and a half, but he explained we could do it that way if we couldn't get a thirteen out of the tree. That day I learned we didn't have to waste that extra half a foot.

This photo was taken in Japan in the early 1970s. Vic and I are in the Emachu office with four of the company's directors.

On my last day in Japan, after visiting all the customers, I went into the hotel restaurant and there was Barry. He smiled and shook my hand. "Did you have a nice trip?" I had to admit that I'd enjoyed it all. After that trip, I got used to travelling through the country on my own.

Over the years, I visited Japan several times. Marilyn came with me a couple of times, and I once went with Vic. We spent six days there, and I got to show him around.

One night, Vic and I met one of our main customers and their family members. The Emachu family had invited us to a bar where they were singing karaoke. The sake was flowing, so I got confident and stepped onto the stage to sing "Sentimental Journey": "Gonna take a sentimental journey, gonna set my heart at ease. Gonna make a sentimental journey to renew old memories." That was the only song I knew. I received rousing applause, but the sake will do that to you.

The Emachu family was great, and Marilyn and I became good friends with them. One of their daughters came to stay with us for a while so she could learn English.

In Japan, Marilyn and I attended a few social functions. We went to several weddings that were modern, not traditional, but we did go to a traditional tea ceremony. Most of the time, the host would ask me to say a few words, and I would say one or two short sentences in Japanese. One time, while we were still at home, I received a warning that on our next trip to Japan I would have to speak Japanese for three minutes. I immediately cornered one of our Japanese customers, who was in Vancouver at the time, and asked him to translate my English into Japanese. He was happy to help and even wrote it out phonetically.

I practised that speech two or three times every day. On the flight over, I asked the Japanese flight attendant to listen to it, and she said I did very well. At the hotel, I read it to the bellhop, and he said it was good, but he added a few words to make it better. I knew they were appropriate words because it would be unheard of to shame a guest. I wanted the speech to be the best it could be to show my thanks to my customers for their cutting and lumber business.

The speech was only three minutes but seemed much longer as I delivered it. Apparently, it was okay because the customers and their wives laughed, especially when I joked that I used to have dark, curly hair when I started in the sawmill business, but after dealing with the Japanese over the years I now had no hair. They liked that one.

The Japanese connection proved helpful bringing in other business. A fellow who worked for the Mitsubishi Company in Seattle once contacted me. One of my customers had recommended S & R, but the man told me he already knew about us because, in Tokyo, the most famous name was S & R: our products filled the docks there.

Marilyn and I are seated in the front row just before I delivered a speech in Japanese to our customers. I look a little nervous.

It felt good to know we had a solid reputation. The man had four booms of yellow cedar to cut, and I lined them up and gave him a price. We did the work, and he was pleased. He gave me a cheque and a bottle of whiskey.

This gift came in handy when one of our main switches broke down on the night shift and we lost power. The big iron push arm had come out of the fuse box and I didn't dare push it in. At 2:00 a.m. I phoned Surrey Electric and waited with the night crew. A guy I knew showed up to fix it. He threw the switch and the lights came on, so we were back in business. He said, "You'll get a big bill for this one, Chick, 'cause it's the middle of the night." I asked him to wait a minute while I went to my office. I retrieved the bottle of whiskey I'd received from the Mitsubishi guy, handed it to the electrician and said, "Thanks for coming out so late." I waited for the bill, but it never came.

Another part of the job is grading lumber. I'm on the right in the sport jacket.

By this time, S & R had established regular routines in the main office that continue today. Tuesday and Thursday mornings we hold a production meeting with our superintendents. We go over the lineup sheets of the cutting schedule to see which company's logs we're going to cut that week. We might have eight sections of yellow cedar going to a specific mill for Monday morning, so we check the lineup with the super of that mill. The customers' names are on the sheets, and they may have four shifts delegated for their cutting, so we organize the shifts and alter them if necessary, moving a couple of guys from one shift to another. The biggest challenge comes when the customer changes the lineup. We cut what they have, but suddenly we've got a space because we don't have all their logs, so we have to take other customers' logs to fill the gap. Some don't have their wood ready, or they want to cut a little more than originally planned.

The process starts with a sales rep from a company like Western Forest Products calling up with an order. Each sales rep handles a species of wood, such as cypress (yellow cedar), fir, hemlock, spruce or red cedar. The guy calls and says he's parked the boom of logs along the Fraser River. If he's got yellow cedar from Vancouver Island, they've been boomed at the logging camp on the salt chuck and could be floating on a bundle boom with big straps around them, or on a log barge, or chained on a flat boom where you can see each individual log. Usually, barges are a better bet on the ocean, so tugboats hook onto them and tow the barges to the booming grounds at the mouth of the Fraser, where their spot's been set by their boom number. The journey could take a week, and if there's bad weather a boom doesn't come in because the ocean is too rough.

When we're ready for the cut, we get the tugs that run along the river to tow the barge right up to our mills. We lease booming grounds along the foreshore. We confirm we have the right boom number for the right mill; for example, in the case of those eight sections of yellow cedar ready to cut on Monday morning, we make sure the logs are going to S mill. The customer always checks the boom, and our head boom man may notice a few logs are missing. The chain may have broken, or if it's rough water the bundle boom can split open and the logs float loose. The log salvage guys gather them up and deliver them to the logging company for a fee. If the logs come loose close to S & R, we send out our boom boat to retrieve them.

S & R is not responsible for the boom until it gets to our booming ground. The booms could have sinkers in them—logs heavy with water—so we get a dredge to bring up the deadheads to protect boats from running into them.

A boom boat gathers logs that came loose from tied bundles seen in foreground. MICHELE CARTER.

All these issues are discussed at the meetings. We also see which type of wood we're doing for the next month and in which mill: we might have red cedar to cut at C mill and yellow cedar going in at A mill. We have to decide which mill is the best for which cut. We might say, "That's not a good cut for C. It'd be better in A." The customer might want A mill because he liked the cut from there, so we agree to cut it there. We're usually two months ahead on our schedule.

During these meetings, other business might come up. We discuss who is in training and how the training's going, who's on vacation, do we have a sawyer for that week, did someone quit, or who's up for retirement after twenty or twenty-five years. At Christmas, we decide who's getting our gift of a twelve-pound frozen turkey that year: we give one out to thirty of our customers as well as to our employees. If we've just hired someone in November, they miss out, but after a few months you're guaranteed a turkey.

TOP Our crews have always gotten along well. Here, I'm delivering a crew talk to some of the A mill and C mill's day shifts.

BOTTOM Turkeys are being transferred to my '71 Chev truck to give to S & R's employees for Christmas in December 1996.

Over the years, the number ranged from 150 to 600, depending on the size of our workforce. We've had a number of employees who received over thirty turkeys before they retired. In 2015, we gave out 494 of them. Any left over, we take to the food bank, and if any of the staff doesn't want theirs, they tell us to donate it there as well.

Twice a year, we get our employees together for a crew meeting. We take two mills at a time to explain what the company's been doing for the past six months, and what we plan to do for the next six months. Some of the staff usually have a few questions for management, so we try to answer them as best we can. We've been lucky over the years. If an employee wasn't suited for the mill he worked in, our preference has been to move him to a different position or mill, but if someone had to be let go I was the one doing the firing. I never liked to fire anyone, but Vic and I would discuss it and make the business decision. We knew it was always a blow to the employee. Fortunately, we didn't have to can too many people.

Vic and I would always keep up communication with the workers. Every day I'd go through the mills and talk to the men when they weren't in the middle of a job, and I'd go in to the saw box and speak to the sawyer. If I couldn't do the rounds, Vic would do them for me. Usually we had customers walking around with us to watch the process and to tell our supers if they had a problem. If any tradesmen came by, I'd check in and see how they were getting on with the work. They were always surprised when I remembered their names.

S & R maintained good relationships with all the trades over the years, and also with the mill workers from our competitors. In the early 1980s, however, we bumped up against the power of an organized union that wanted to make inroads into S & R, but I wasn't in a welcoming mood.

8

UNION COMES CALLING

HEN A COMPANY decides to build a sawmill from the ground up, the owners need to know what they're doing. The other ingredient is luck. One of our competitors cut three-by-eights and three-by-tens for Europe, but they didn't want to cut big timbers. At that time, big logs were still coming out of the bush, so we got the call to take those cutting orders, and we were ready. This was a lucky break for us. No one can predict the market. In 1980, we got B mill running two shifts a day cutting fir and hemlock.

As business rolled our way, we needed to expand our operation. We purchased fifteen acres alongside B mill, built a maintenance building with a machine shop and garage, and hired around twenty-five more guys. In 1985, we acquired another mill on fifteen acres east of the shop, across the creek in Langley, and set this up as D mill. We knew the guys who had been the previous owners because we'd helped them out when a union was thinking about moving in.

Originally, in early 1981, a group of fifteen guys pooled their savings to build the mill. They had good careers in sawmilling and machinery; most came from Field Sawmills in Courtenay, BC. They planned, designed, arranged financing and insurance, purchased land, got zoning and approvals, bought used and new machinery, and started construction work on the fifteen-acre site in March of '81. These were hard-working guys, pulling in long hours six days a week until the mill was ready to saw its first logs in early September. The whole operation had been completed in just six months. They called it West Langley Forest Products. The mill started running two shifts in November, and the guys were seeing profits so they felt optimistic about the business, but they got nervous when the union showed up.

The union guys came over from Vancouver Island to try to organize the workers at West Langley. The company employed about eighteen men and knew who was in favour of the union. The numbers weren't good, so we sent two of our guys over to work there and get included in the vote. The union lost by one vote, so it never got in. Union guys used to call Ray Chretien, who was B mill's foreman at that time, and they'd say, "Ray, what's Chick doing? Nobody wants to sign up." Our employees liked working for us; they wanted no part of a union.

We always appreciate our workers. My philosophy is to treat others the way I like to be treated. I'm firm but fair. S & R started a pension plan so our employees have a nice little pension when they retire. We have family medical and dental, so if a guy is thinking about quitting, his wife will set him straight about the advantages of staying covered. We also include eyeglasses in their medical coverage, something the union didn't do, and if any worker becomes sick or has a minor injury like a hurt finger, they can do their rehabilitation in another section of the mill until they get better, so they never lose full pay. We stay in line with the average sawmill wages and give workers increases along the way.

Between 1973 and 1992, Jack Munro was the head of the International Woodworkers of America (IWA), the forest industry union.

He was a prairie guy like me, from Alberta, and he put his all energy into the trade union movement. He had the power to take thirty thousand workers off the job and start a general strike, a hard blow that terrified politicians. He once told me, "Chick, I'll get you unionized, no doubt about that." I told him, "You won't live long enough." He was a good guy. Just doing his job for the members. He wanted to make trade unions an accepted part of society and make sure the members were treated fairly. I had no disagreement with that.

In the 1970s and '80s, forestry was BC's largest employer. Jack's union had forty thousand members during those years, and he negotiated down to the last dime for the workers. His position changed in the '90s, and in 2004 the union merged with the United Steelworkers of America. By then, the BC membership had slipped to twenty-eight thousand because of layoffs and several mills closing. Jack was tough and persistent, but he never got a union into S & R.

That didn't stop him from picketing us. In 1980, we had two hundred striking members blocking production at S & R. Jack's members didn't want the non-union mills to keep running. The IWA had been striking for five weeks before signing an agreement with the forest industry. When our employees came to work on a Friday morning, they found the pickets had jammed our gates and wouldn't let our guys enter the three mills. Only management was allowed to go through. During the whole day, they kept intercepting vehicles, saying S & R was custom cutting lumber for Crown Zeller-bach, a forest company that employed nine hundred striking IWA members. The strikers wanted S & R to stop cutting Crown's wood. The union guys got in a boat and spray-painted the word *hot* on our booms to stop us cutting them. I told them we were cutting an old order. When the IWA strike began, we told our purchasing manager not to accept any new orders out of the "hot" booming grounds. We stuck to our promise.

One Friday night, I was still at the mill. The night shift guys were showing up, and the picketers were down to six men. I drove my truck over to them and we had a good talk. A couple of them knew me because their fathers had worked for me at MacDonald Cedar in Fort Langley. My workers were in a lineup waiting to go in for their second shift, and the union guys waved them through. We understood the strikers' position and respected it; we wouldn't cross their lines, so that night I was relieved when they let our employees go through. Fortunately, the deal was signed that weekend, so they didn't come back on Monday morning. The pickets told us they were still planning to come back—not to strike but to organize our three hundred employees and bring them into the IWA. Didn't happen.

One time, the shake mill next to us had about fourteen union employees go on strike. We were building a filing room in A mill at the time, so it was shut down for repairs. Pickets were blocking the roadway that served the shake mill and our mill, and we wouldn't cross the line, so we came up with another plan. The filing room didn't have a roof yet, so we hired a helicopter to haul the timbers overhead and lower the roof beams from above. The guys on the picket line didn't get angry: I think they admired our ingenuity.

Another one of our neighbours, Ken Mackenzie, owned Harken Towing, a tugboat company that towed log booms to sawmills on the Fraser River. Ken had been around since the late 1940s, and had built his business on putting his customers' interests first. One year his crews went on strike, which meant S & R wouldn't get the booms in place. Ken was a good friend who wanted to keep our business going, so he got in one of his tugs and, with the help of our boom boats, hauled logs right to our mills.

One morning, his wife Dorothy came out to help. She had little experience with the tugs, but Ken needed her to steer. He jumped onto a boom stick to chain a boom of a few sixty-foot logs to the tug while Dorothy waited in the wheelhouse. Before he could connect the chain, Dorothy put the boat in gear and started moving away from him. He was yelling at her to back up: "Put it in reverse. Reverse!" She had no clue how to do that, so she tried to steer around and head for him as he floated down the river on the boom. We used to joke that when you couldn't get hold of Ken, he was still stranded on the boom, floating out to sea. "Too bad, 'cause he was a nice guy."

Like Ken, we looked after our customers and our employees as much as Jack Munro did his members. Throughout the years running S & R, I was continually reminded that if you choose the right people to do the job, they'll make the company work. From the start, Vic, Marilyn and I invested in our workers. Until a few years ago, I would hand out the paycheques to each employee on every shift. I knew all their names, and I liked to have this brief exchange because I could keep up-to-date with any problems or hear about someone becoming a parent. If it was good news to be celebrated, I'd pass the information on to Marilyn and she'd organize a card or a baby's gift. If a superintendent told me about a worker not doing so well, I'd be able to tell him to lighten up on his labour because he was going through a rough patch, and the next time I gave out the cheques, I could speak to the worker and ask how he was doing.

I remember one winter getting a call at the office from a former employee. After working for us hauling sawdust, he had to move out of town and was calling to say hello. He was up country on a logging road, ploughing snow and had been thinking of S & R. "It's lonely out here on my own," he said. "I haven't seen a soul all day, and I started thinking of my time working for you. I was kind of missing you,

Chick, and just wanted to hear how you're doing." We had a good talk and he said it was nice to hear a familiar voice.

As I mentioned, we pay our employees a good wage, but we also keep work schedules flexible to meet specific needs of our customers. Our employees form a loyal, highly skilled and highly motivated workforce, and we let them know we appreciate them. The union couldn't compete with that. Our workers always rejected the unions through democratic vote.

We had also helped keep the union out of the West Langley mill, and a couple of years later the fifteen owners came to us with a new request. They needed to sell, so in 1985 we bought it from them. They had known there was risk in any business, but they hadn't expected that the lumber markets would let them down. By late 1982 and into 1983, log supplies became tight and lumber markets dropped, while interest rates rose to 20 percent. The downturn forced permanent mill closures and up to twenty thousand industry jobs disappeared. The owners didn't have the personal resources to put in more money, and the bank wouldn't increase their loan, so they were forced to close the mill. It sat idle as the owners struggled to carry the loan. The markets would take another eighteen months to improve, but during that whole time they would be stuck with paying their loan. If they couldn't find a buyer, they'd go bankrupt trying to carry the loan payments with no production going on in their mill.

S & R saw potential in the newly built mill, right next door, so I went to the Royal Bank to negotiate. I made an offer, but the bank manager refused. Back at the office, Vic and I were discussing raising our offer when the bank manager called. He said the manager in Toronto wanted to get it off the books and told him, "Call that guy back and sell him the damn mill!" The owners saw the sale as a positive solution to their problem. A few of the guys came over and worked for us, and they're still working at S & R. We named this new property D mill, and it's been a very good production mill for cutting different species.

By the mid-1980s, our daughters were pitching in and helping Marilyn in the office. We had also built a new office to replace the original one from years earlier. That old one is still on the property.

Our first mill was long gone, and we built a new A mill in its place.

By 1986, the business was doing well. We were cutting eighty-four million board feet of wood annually, mostly going to Japan. At that time, we cut 65 percent of the total yellow cedar and spruce markets. The Japanese business could come in cycles: when they got low inventory, they'd start buying and the price of logs would shoot up, but then their places over there would fill up with lumber, so they'd put a hold on production. Our lumber went to small-scale sawmills in Japan to be remanufactured into many different sizes. "Reman" mills, we called them.

We were doing work for about twenty companies. Our regular customers included B.C. Forest Products, Crown Forest, MacMillan Bloedel and International Forest Products. During this time, the senior people at these companies were men, and every year Vic and I would attend a charity function for the BC Children's Hospital at the Vancouver Club with these lumber guys. Occasionally, we took along our top employees to put faces to the people they'd been dealing with all year. We usually sold three hundred to four hundred tickets and raised a good amount for the charity but, to make the evening more entertaining, we also had a raffle with each of our names in a box. The last name left in the box would win $10,000.

Over a few hours, while the names were being pulled, about twenty of us had a little side bet. We put our names on a sheet of paper and bet $20 we each would outlast the others in the raffle.

TOP S & R's office.

BOTTOM S & R's office in winter. We never could part with our first
lumber carrier.

S & R's A mill with its newly completed filing room had long ago replaced our original mill.

I'd go around collecting the money. One evening, about five names remained in the raffle box: mine had already been called. I noticed that the CEO of Interfor, Bob Sitter, still had his ticket.

"You might pull this off, Bob."

His head was drooping in his hands, and he couldn't keep his eyes open.

"I'm not going to make it, Chick. It's after one o'clock and I'm beat."

If a man left the party before his name was called, he forfeited his winnings, so I could see Bob was in a bit of a bind.

"How about I buy your ticket off you?"

Bob's eyes opened. I could tell he really wanted to get home.

"How much?" he asked.

"Five hundred dollars."

"Double that and you've got a deal."

I gave him $1,000 and he handed me his ticket. Another half hour went by before the final name was pulled from the raffle. "Bob Sitter!"

That night I made $9,000.

By 1992, S & R was employing about five hundred workers and business was going strong. We were cutting 190 million board feet of lumber, all of which was being shipped to the Far East. Of that, 100 million board feet was top-grade yellow cedar, and a single forty-foot log was worth up to $10,000 (or the amount of a raffle prize). The sawmills were running, and the lumber companies were successful enterprises. No one can forecast the future, but we felt the industry and our markets were in good shape—until a shock wave hit the coastal lumber industry.

9

FIVE MILLS ON
THE FRASER

BY EARLY 1997, our four mills were busy day and night with ten shifts operating in A, B, C and D mills every day. The custom cuts were converting selected logs into lumber products specified by our customers, so we could've called it "customer cutting." Containers were shipping our lumber to several ports in the Pacific Rim where it continued to be remanufactured. Business was good for S & R, but the BC forest industry had a bad 1996, losing about $250 million, and was facing an even worse 1997.

On the coast, about 50 percent of wood products were going to Japan and about 18 percent were shipped to the United States. Japan was a major customer, so when the Japanese economy tanked during the Asian financial crisis, BC's coastal forestry got hit hard. House construction slowed down, and the coastal companies lost a major outlet for their lumber. They suffered the worst slump in fifteen years and had to throw five thousand people out of work. I knew the guys in these industries, and they figured in the first quarter they were going to lose around $200 million in provincial revenue as sawmills and logging camps shut down. Interfor's CEO Bob Sitter and I used to chuckle over the $10,000 raffle ticket I'd bought from him, but at an industry event neither of us were doing much laughing. He told me, "We're in a box. We can't break out of it. The only option we have is to wait until Japan recovers." The timber harvest went down from 40 to 60 percent that year. S & R managed to weather this storm, even though we had to drop some shifts, and in 1999 the Asian economy began its recovery.

The lumber industry had other problems. Groups were concerned about the environment and especially logging going on in the old-growth forests. Forest companies, First Nations, environmental groups, government, local communities and loggers were all fighting it out. They finally figured they needed to work together and compromise if anything was going to change.

S & R wasn't directly affected, but we had our own challenges with log exports. During the 1980s, house construction was good: we had a building boom, so lumber was manufactured locally and exports were small. When the coastal forest industry does well, log exports go down, but the opposite is also true, and the forest industry suffered in the 1990s. Foreign restrictions were killing the lumber companies: they had everything thrown at them. The United States imposed import duties, tariffs, quota restrictions, whatever you want to call them, and all these foreign barriers caused reductions in log prices. The Canadian dollar was going up and down; the Japanese market for hemlock collapsed; the change from harvesting old-growth to second-growth timber made for a bumpy ride, and from 1996 to 2006, more than twenty-five mills closed. For years, the government had made it a law that before a company exported its logs, the sawmill guys had first dibs on purchasing them. We blocked the export, bought the logs, and carried on business, but we occasionally had to cut shifts. The economic circumstances affected us, but S & R was never in such bad straits that we had to close down our operation.

In fact, business was doing well enough that by 2000 we were ready to build another mill. On the east side of our property, we built S mill as a specialty cut mill. We positioned it near our machine shop on river frontage. By this time, we were cutting all species and sizes. We were also shipping to the United States. Our customers were bringing in red and yellow cedar logs from all over Haida Gwaii on the north coast of BC, Douglas fir, hemlock, spruce and pine from Vancouver Island, and the local alder for furniture. In S mill we still run one or two shifts, depending on demand. We cut mostly cypress, fir and hemlock there, and can cut long forty-foot logs. We also had one merchandising log mill, a planer mill and a container yard. Our fabrication shop is staffed by twenty men who repair our machinery and build parts necessary to keep S & R functioning constantly.

In 2002, we employed over five hundred workers on three shifts in a twenty-four-hour day and were cutting 190 million board feet annually. Of that lumber, 42 percent was cypress (yellow cedar). Unlike larger mills like Canadian White Pine and Fraser Mills that had shut down in the previous years, our operation was shipping 90 percent of our output to Japan and other parts of Asia, specifically Taiwan and Korea. We saw increased competition from others selling to Japan at this time, but they weren't as much into custom sawing.

On the western edge of our property is E mill, a whole-log chipping plant that employs eight men. The chipping plant on its own is a good little business. We added four chip-loading dumpers, and in 2002, about 140 chip trucks came into our area per week to dump chips on our automatic dumpers, and we were loading approximately twenty-five scows (floating chip barges) of chips, sawdust and hog fuel per week. In the present day—2017—an average of about a hundred trucks a week drive onto the site carrying chips to be dumped on the automatic dumpers.

LEFT A chip-truck dumper. The trucks that deliver the chips come from up country.

RIGHT Chip conveyor at E mill. On the right, the chips are dumped into a scow in the river.

Chip trucks come from all over the interior of BC with different species of wood chips. We keep the species separate in each pile because they produce a different quality of paper. The bark mulch becomes hog fuel that goes into the pulp mills to feed their boilers. The chips, sawdust and hog fuel run along conveyor belts to the chutes, where dumpers stockpile the loads onto scows, a process that takes about six hours. Catalyst Paper in Richmond takes most of the chips, and Harmac Pacific on Vancouver Island takes some. In the old days, the forest industry would leave smaller logs in the bush, but now the junker trees are chipped. Nothing is wasted. Our customers own the logs, so our only stockpiles are chips and sawdust.

As a nod to my dad, our chicken farmers can always rely on a supply of sawdust. At one time, we politely turned down a giant lumber company that requested the farmers' pile as well as their own. I told the company, "I can't let the farmers down. I gave them my word and that's all there is to it." They didn't argue. They knew I wouldn't break my word.

After so many years in the business, I had an eye for seeing a junker log hiding inside a boom. Around the time we were building S mill, I got a call from Frank at Canadian Forest Products to go out to the booming grounds by the University of British Columbia. He said, "I've got two booms of pulp cypress." I asked him to give me the CFP boom numbers and I'd go take a look. I went by boat with one of my guys to see what he had for sale.

The booms were bundled with cables, so I decided to take a closer look at the product. I put on my cork boots and started walking over the logs. I noticed they'd put the CFP tag and number on their best log, a common practice. The idea is the salesman will say, "They're all like that." I wasn't convinced.

I kept walking over the boom and then I saw it—a real junker. I turned around and got back onto the boat to change into my overalls and grab a rake. This log was in such bad shape it took no effort to start hollowing it out. I kept going until I could see light at the other end. I phoned Frank and told him what I thought of his logs.

"It's a bunch of crap here."

"What do you mean? Those are great logs."

"You got a log here that if I stand on it, I'll squash it."

"That's crazy."

"What's crazy is why they didn't leave this in the bush. It's the dirtiest log I've ever seen. I'm still checking to see how many others are in terrible shape."

We stayed on the phone while I negotiated the price. I ended up buying the boom but for a much cheaper price than he had wanted.

As a joke, I took the CFP tag off the prime log and nailed it on the rotten log. I crawled inside and the guy I was with took a picture of me sticking my head out. My whole body was inside that hollow log. I sent Frank the photo and he got a chuckle out of it. He sent the

Here I am inside a hollow log—a real junker!

picture to the logging camps, and they made copies and hung it on their bulletin boards. No explanation required!

For the next few years, business steadily increased. Customers' logs were lining up along the river frontage by our mills, so we purchased waterfront property across the river on Barnston Island where we could store log booms waiting to be cut.

An overhead shot of our five mills on the Fraser River.

The last property we bought to complete the S & R family of mills was located near C mill. Lloyd Brown's Fraser Pulp Chips, a sawmill specializing in low-cost western red cedar, had been operating since 1962. In October of 2007, his mill went on temporary shutdown when the log market dried up. He told me, "It's harder for us to get any supply. And what there is out there, the price is through the roof." In February of 2008, he sold the business to S & R, which brought our work site up to two hundred acres.

S & R never had a sales force, so over the years our business grew mostly from word of mouth and from our reputation for making customer satisfaction our priority. We maintain basic price levels when markets are strong or weak to keep our regular customers. They generate enough work in the slow periods to keep our mills running and our crews working. Sometimes, we make a little concession here and there to keep the guys working, rather than lay them off. This is a fluctuating industry, and we try to maintain certain standards in order to prevail: we don't inflate prices when the going is easy. Normally, if it's a good market, our regular customers want to increase their cutting. We give them that benefit because they've kept us going in the slower periods. As the profits increased, we were able to inject millions of dollars into upgrades and improvements.

We also try to keep our mills safe by focusing on training and responsible work habits. Our attitude toward safety is just as sensible as our fire protection concerns. If we were to have too many accidents, the compensation would hurt our business. Our low injury costs earned our company a big discount on the Workers' Compensation Board's rating system, so we pay much less than the base rate for insurance. That discount translates into hundreds of thousands of dollars a year, which is money we can invest back into our business, our employees and our equipment. Workplace safety in a sawmill is more than our policy: our safety meetings and ongoing training keep prevention of accidents at the front of everyone's mind. The lower the injury costs, the less you pay. This is just common sense!

Fifty years after we bought the first mill, we invested in new Cat wheel loaders. We were working with 240 different sorts of lumber of different lengths, thicknesses and grades, and needed to keep production moving. Equipment breakdowns cost money and production time so we figured we'd better invest in newer technology and try to cut fuel costs. We've got close to 150 pieces of equipment from forklifts to wheel loaders to boom boats. We can run ten wheel loaders that do everything from bucket work moving chips to separating the log booms—that's when we use a grapple-equipped loader. The customer still comes in and looks at each individual log to identify what kind of yield and product he wants out of that log. We put the log through the mill and cut each one specifically to his directions just like we did in the beginning.

For years, we'd figured out how to run a lean operation. We worked on used mobile equipment that we'd pick up at auctions, and we'd run it forever. The practice of hanging on to our older equipment was based on cost, but were we saving by not purchasing new equipment? When equipment keeps breaking down and we lose production because of it, the savings disappear, especially if we spend too much time and effort repairing the older stuff. We had a spare loader in case the main loaders broke down, but what did we do when the spare went down and we had to fix it too? Our only option was to rent a wheel loader at a big cost. Our crew of mechanics is top notch, but they were practically fixing the machines with string and duct tape. Time to buy new and upgrade the equipment in our fleet.

Even with these improvements, we still see ourselves as an old-fashioned sawmill, where you take the log and turn it. It's like the butcher cutting up a side of beef: you've got to get the sirloin steaks out of there first, and then keep going right down to the hamburger; you cut that log like you're going to get the best choice cuts out of it.

One thing a visitor to our mill site will notice is that the buildings are all painted green. My father was from Ireland, and emerald green is my favourite colour. One time, an employee painted an edger for C mill an orangey-red colour because he thought it looked good. I asked him to get the can of green paint and give it another try. The buildings are green; the walls are green; the catwalks and handrails are green; and the beams are mostly green. I had to let the guys paint some of the beams yellow because they kept bumping into them in the darker sections of the mill.

My fully restored original '71 Chev. WAYNE LEIDENFROST.

My 1971 Chev pickup was the first of its kind to have air condi-
tioning. A special feature, but I made that truck even better when I
painted it emerald green. It still runs to this day.

In 2003, just before my seventy-fifth birthday, and fifty years
after Marilyn and I had first met, we thought it would be fun to go
back to Fort Langley and see if the ice cream parlour where it all
started for us was still there. The place had been renovated but the
ice cream was still good.

In 2008, I was going to have my eightieth birthday. A long-time
customer came by and reminded me about that locked safe that was
still sitting on the back porch of our original office, and I thought
now might be the right time to take a look inside. We had built a new
office and forgotten about the safe until our customer asked, "Did
you ever open that safe?" When I told him we hadn't, he said, "After
all these years, that safe is yours. Why don't you satisfy everyone's

Marilyn and I return to the ice cream parlour where we first met.

curiosity and open the damn thing?" I spoke to Marilyn about this, and she agreed we should take a look.

We decided to open the safe at our house on my birthday when the family would all be there. I phoned a locksmith to meet us at the house, and got a forklift to put it onto my pickup truck. The four-foot high safe probably weighed over a ton. I drove over to our house and had one of the guys put the forklift on another truck and follow me. The family gathered around as we watched the forklift driver take it off my truck and place it on the driveway. The locksmith took his bag of tools and knelt in front of it to listen to the combination lock's tumblers.

My family, like the employees and customers, had their own ideas about what might be inside. The most popular suggestion was that the guy who owned the safe was planning to pay for Harold Morrison's mill with cash he'd stowed inside. "He got in a car accident, or he had to leave the country because he'd got the money illegally, and the cops were on his tail."

"If it's gold bars, they'll make a nice birthday present," joked Marilyn.

I told her that even though it was *my* birthday, she could have the gold bars.

"Maybe it's important documents."

"Would he leave Deeds of Land in there?"

"He'd have come back for those."

"Not if he's dead."

"Or had to leave the country."

The locksmith finally found the right combination and stood. "It's all yours."

I wanted everyone to be able to view the opening of the safe, so we got the forklift off the other truck and positioned it under the safe to lift it back onto my pickup. I stepped up into the truck bed and rubbed my hands together like they do in the movies before they crack open a locked safe. My family egged me on. I squatted down, bent my ear low to the combination lock and carefully put my hand on the lever. I pulled it down and opened the door.

"What's in there?"

"Let's see."

"Tell us! We can't see."

The contents of the safe had left me speechless. I suppose I had a few ideas of my own floating in my head: could be a key to a safety deposit box, the guy's will, or maybe a pack of letters. After all those years of guessing, it came down to this. I put my hand inside and retrieved a piece of paper with some ink marks on it and held it in the air.

"Is that it?" they yelled.

Is that all there is? After waiting 45 years to open it, we found no gold bars or treasure map inside the old safe.

I handed it to Marilyn. Something about that sad piece of paper struck us funny and we started to laugh. Didn't take long for the news to travel around the mill that Chick had finally opened the safe, and the "damn thing" was empty.

THROUGHOUT THE YEARS, my family has taken an active role in the business. My sister Margaret was the exception. She never worked at the mill. Marg married young and remained a home-maker all her life. I didn't see much of her over the years; when we were still all at home with Mom and Dad, she spent most of her

Me and my siblings at a restaurant in Maple Ridge in the early 2000s. Back, left to right: Denny, Sam, Herb and me. Front: Dave and Marg.

time with Mom and never got involved with her brothers' pranks and adventures.

Marilyn was always front and centre in the office; my daughters Wendy, Suzanne and Colleen would spend their weekends with her during their school years. Wendy and her husband Barry worked at the mill for over twenty years; Colleen's husband Dave has his own trucking company and poultry farm; Suzanne's son Jeffrey is a ticketed millwright at the mill; and her son-in-law James works in the container loading. Her two daughters Jaclyn and Michelle and Colleen's daughter Jennifer work part-time in the office while Suzanne and Colleen have taken over Marilyn's administration duties. Colleen's son Jeremy is an apprentice electrician who is also working at the mill, and her eldest son, David, started at the mill as a clean-up boy and then worked on the green chain for several years.

Three of my four brothers and my dad also worked for me in the sawmills. My youngest brother Herb was the first to come to S & R in the mid-1960s. He worked in the yard for many years shipping lumber. Herb now lives nearby in Maple Ridge and recently retired after working one day a week at the mill as a watchman. My older brother Dave had worked at Hammond Cedar in the maintenance shop and came over to S & R for a couple of months in the early days, but he was busy with his twenty-acre farm in Pitt Meadows where he boarded race horses, so he didn't stay at the mill. He became an invalid for about ten years, confined to a wheelchair, so I gave him a golf cart, which made it easier for him to get around the farm. He died in Pitt Meadows on his farm in 2015.

Sam started with S & R in 1963 and stayed about thirty years. He worked in the shop and on the truck dumpers, where he talked to the truck drivers and checked the dumpers. He later retired to Kamloops, where he died in 2015. Like Dave, my brother Denny also worked at Hammond but in the boiler room. He stayed there his entire working life. Denny died in 2012 in Maple Ridge. In 2013, my sister Marg died from pneumonia in Langley Memorial Hospital at the age of eighty. My dad and mom had moved to Fort Langley near the house where I had boarded on Wright Road so, after working in a little sawmill in Haney, Dad eventually came over to work for me as a watchman until he turned sixty-five. He got sick and passed away in the hospital in Langley in 1969 at the age of sixty-nine. My mother lived for a few years in an apartment in Maple Ridge before passing away at the age of seventy-six in 1981. Of the six siblings, Herb and I are the only surviving Stewart kids.

In 2013, Marilyn and I celebrated S & R's fiftieth anniversary. We'd both worked hard over the years and had been lucky with

TOP S & R's 50th-year celebration.

BOTTOM Chad Day, me, Marilyn and Geordie Vandekerckhove. Chad and Geordie are two of the top guys at S & R.

In the mid-1960s, a young Vic Rempel caught this beautiful steelhead on the Babine River.

attracting so many good workers to help us keep the business going strong, so we decided to have a little party for past and present workers. We got some gift bags and put in each one a Stewart family pen, a $50 bill, an S & R hat, a cookie iced with our company logo and a key chain/flashlight. We met with all the employees and took a picture with them and thanked each one of them for their hard work and dedication. We had photos on the wall of employees who had worked for us over the years, so everyone got to identify people and share some memories. Then we had a lunch of deli buns and chips and a beverage. We all enjoyed the day.

In 1986, Vic Rempel sold out to me and retired. Like me, he'd never had an accident in the mill or lost any fingers. His two sons, Gary and Bob, came into the mill full-time after high school, doing cleanup at first and landing better jobs over the years. Both sons recently retired from S & R, but dropped by the mill this year for coffee. They reminded me about their dad being such an honest guy. The story went that the family was eating out at a fancy restaurant and Vic was $5 short, so he couldn't cover the whole bill. He gave the manager his word he'd be back with the money. He drove the family home and went back to the restaurant and paid the manager the five bucks. He said, "I wouldn't be able to look myself in the mirror if I hadn't paid up." He was an ethical guy with common sense, and that made him a good partner and friend.

He was also an excellent steelhead fisherman and, before he left S & R, we once fished the Babine River together.

The Babine in north central British Columbia is a tributary of the Skeena River and became a favourite retreat of ours. Sadly, Vic passed away in 2010, and because he loved the Babine his family spread his ashes over the river.

In the early days of S & R, Vic and I probably would have gone back to the Babine to fish together a lot more, but an incident happened at the mill that made it impossible.

10

GONE FISHING

Y MOM USED to say, "All work and no play makes Jack a dull boy." She figured if all you had to talk about was your job, you were probably a boring person. I love my job but, with Marilyn's encouragement, I learned it's important to take breaks to enjoy life. For us, vacationing out of town did us the most good.

The first getaway property we owned was on Gossip Island. We'd bought two lots in the 1950s. As I mentioned earlier, Gossip is one of the Gulf Islands in the Strait of Georgia and sits at the southern tip of Galiano Island. Active Pass is the narrow passage that separates Galiano from Mayne Island, and is one of the routes BC Ferries uses to get from the mainland to Vancouver Island. We can often see the ferries on their runs through the pass, but no ferry comes to Gossip.

One day, Marilyn and I decided we should build a yellow cedar cabin for our island retreat. We had the cabin built at B mill and put it on a barge to ship to the island.

We travelled to the island by boat every weekend in spring and summer. On one of our trips, we were towing our boat on the back of my '71 Chevy pickup as we drove to Crescent Beach to launch the boat. As usual, Marilyn, my three small daughters and I were jammed into the cab of the truck so tight that one of our daughters had to sit in the middle with her legs on either side of the gearshift. We lived in Coquitlam at the time, so we had to drive over the Port Mann Bridge and up the steep Scott Road hill.

Everything was going well until the truck stalled at the crest of the hill. Shifting gears was difficult as it was, and now the truck and the boat trailer were rolling back. I put on the emergency brake and told the girls to stay calm as I tried to figure out how to keep one foot on the clutch, one foot on the brake and one foot on the gas pedal. I needed three legs! I'm not sure how she did it, but while the girls scrunched up to give her room, Marilyn managed to lift her left leg over the gear shift and stretch as far as she could to get her foot on the brake pedal. I slowly let the emergency brake go and put the truck in gear, dropped the clutch as Marilyn's foot came off the brake, and gunned the gas with my right foot. The truck jerked forward and Marilyn untangled herself as the girls started breathing again. We made it up and over the hill and had a good laugh at our team effort.

Our trips to Gossip often involved adventure, especially when we motored across the strait in our boat. The journey took forty-five minutes and on a good day it was a pleasant ride, but when the weather switched and the waves grew, we all secretly wondered if we were going to die. We once saw a dead baby whale floating on the water and were scared it would tip us over.

Another time on our way to Gossip, we saw a ferry stalled in Active Pass, with several members of the crew lowering a rescue boat. We saw them lift a bloated body from the water and put it in a net to lift onto the ship. We found out later that someone on board had spotted the dead body so the ferry had to stop: a man had committed suicide by jumping overboard.

TOP We built the cabin at B mill, then loaded it onto a barge bound for Gossip Island.

BOTTOM The cabin being towed toward Active Pass.

Our yellow cedar cabin settled in at Gossip Island.

One terrible tragedy we witnessed in the summer of 1970 involved a Russian freighter ramming into a BC ferry in Active Pass. We heard the squeal of the ships trying to break to avoid each other, but the freighter crashed into the ferry and almost cut it in half. Three people died. The ships had tried to contact each other before they collided but weren't on the same radio frequency. We watched small vessels motoring to the scene from Galiano Island and could see the ferry passengers with life jackets on, leaning over the side of the deck, waiting to be rescued. The freighter wasn't supposed to be in Active Pass, so the Russian government had to compensate BC Ferries.

Another time, with my family on board, we had a problem with the boat. I was at the helm, steering through Active Pass when Suzanne told me we were taking in water. I had my gumboots on, so I went back to see what was happening. Sure enough, water was coming in, so we put the lifejackets on the girls and I pulled the plug to let the water out. I could see a crack in the bow and managed to get over to Gossip, but I wondered how we were going to get the boat back over to the mainland. The next day as we were getting ready to go home, I put a tarp over the bow so that when the boat motored against the wind, the tarp would get sucked into the sides and shut out incoming water. The plan worked long enough for me to get my family onto land, but I still had to take the boat over to where I had parked my truck.

I was on my own, steering the boat and praying I'd make it safely to shore. The tarp had worked well for most of the journey, but it was finally done; I could see water seeping into the boat. Two guys were fishing in a small boat a little ways ahead of me. I yelled over to them, "Can you give me a hand? I'm taking in water."

They brought their boat over and one of the guys jumped into mine.

"You steer," I said. "I'll try to do something in the back."

I started bailing like crazy, and the guy steering yelled that we were almost at the shore. We had made it to Crescent Beach without sinking. That day, I was sure glad to step on dry land. The two guys helped me get the boat to my truck and put it on my trailer. That was quite an adventure.

When Marilyn, our three girls and I went over to Gossip, we roughed it. We had no electricity or indoor plumbing, and we mostly slept in a tent. We had constructed a little tin house where Marilyn cooked in the makeshift kitchen that had a big piece of driftwood for a countertop. She would be dressed in her jeans, a navy blue sweatshirt and white bandana over her blond hair, holding a frying pan sizzling with fresh bacon. We all loved eating outdoors with the smell of the briny ocean and the campfire smoke mixing with Marilyn's cooking.

Marilyn and I are enjoying a relaxing afternoon on Gossip Island. Note Mt. Baker behind the ferry.

I always liked to fish, so I wanted my girls to enjoy the sport too. We needed to fish according to the tide, which meant waking the girls early to get in the boat and go jigging for herring. Herring are filter feeders that swim with their mouths open through plankton, so our chances were usually pretty good of hooking those open mouths. The girls would be half asleep, wrapped in blankets and holding their poles, but would wake up pretty fast when they got a bite.

When the tide was right, we'd see thirty or forty other boats with one or two people in them fishing in Active Pass. One Sunday morning, I was a few feet from another boat and had two lines out. I got a catch and was trying to bring it in when the other line went. I didn't want to lose the rod, so I called over to my neighbour, "Take this line. It's got a fish on it." He took the rod, and I followed the other catch that was taking me out a fair ways before I could land it. When

I returned to the other boater, he handed me the line and the fish.

"You keep the fish," I said. "I've got this one. I just need my fishing rod."

He was happy to oblige.

We never had TV or a radio in our cabin, but we had plenty to entertain us. We'd play cribbage or horseshoes, and the kids would do arts and crafts. Marilyn would keep the camp tidy, and I'd help prepare fish for dinner or shuck oysters when they used to be out on the point. When the oysters were still around, Marilyn's cracker crumb oyster recipe was a real treat. We can't have campfires anymore, but food still tastes great when we eat out in the fresh air.

Roughing it meant we learned to make do. We always brought in our drinking water and, over time, we set up the well to have clean water, hooked up the electricity and installed toilets, but we still used the ocean as our bathtub. Marilyn would give the girls their evening baths out in the grooves on the heated rocks. They'd wait until high tide, almost before bed when the sun was going down, and she'd pour warm water on them. Not as convenient as home, but we liked that about the place.

When our daughters had children of their own, we had one medical emergency. Suzanne's son Jeffrey split his head open on a stump. He was about five years old and was on the homemade swing, which was just a buoy we'd attached to a rope hanging from a tree. Someone had given him a push and he lost his grip on the rope and fell off. He hit the jagged side of a stump and split his head. Suzanne grabbed a towel and kept it against the wound while we took our boat to Galiano and then to the ferry. We called the hospital and got a water taxi to take us all the way to Ganges to the hospital. Jeffrey got his head stitched up and was able to go home right away.

Over the years, we were lucky not to have many injuries on Gossip. Marilyn's allergy to bee stings kept her prepared with her

I have a successful early-morning fish near Gossip Island with our daughters Suzanne and Colleen (in the red sweater).

EpiPen. These days, if we ever have another emergency, we'd just call 911 for a hovercraft to come to the island and take us directly to the hospital.

In the summer, after the kids were grown, Marilyn and I would go over to Gossip on our own right up to when I was in my early eighties. To keep our daughters happy, we always wore our survival suits in case we fell in the drink. We'd get into our 200-horsepower Boston Whaler and leave Crescent Beach at 7:00 on a Friday night in time to get to our cabin just before dark. We'd get to the dock on

TOP Fishing in the fall has been a favourite pastime of mine.

BOTTOM Marilyn takes in some sun on a rock out in front of our cabin at Gossip Island.

Marilyn and I always looked forward to our time off work. Here, we're motoring near Gossip.

Gossip, tie up the boat, carry in our food, and stay for the weekend. On our way back home, Marilyn always packed a little picnic for us to have midway. We'd turn off the engine, open a bottle of wine, and eat cheese and crackers while we floated and enjoyed the day.

The island is still a fun destination for the family. In 2016, we celebrated our fiftieth annual Little Gossip Island Fishing Derby. Part of the celebration included me cutting the cake because it had been fifty years since Marilyn and I bought property on this piece of paradise. I also happened to be the oldest Gossip Islander at the derby.

One of our neighbours donated a big field for the derby party, and a few hundred people showed up—twenty-five were from our family. The island has stayed pretty much like those early days in 1948 when I got the mill job oiling and greasing machinery for the

Hartnell family. I never imagined that one day I'd be buying property on Gossip or that the island would become a great retreat. I was lucky on that one. Another place we loved was inland, away from the ocean, and near where Marilyn used to live.

When we were first married, we used to visit our friends' fishing lodge on Babine Lake north of Smithers in central BC. Babine Lake, at 110 miles long, is the longest natural lake in BC. Fishing is great there because of the rainbow trout and sockeye salmon. The fish make their way to the north end of the lake where it narrows into the Babine River, a sixty-mile stretch of water that flows west to the big Skeena River system where anglers fish wild steelhead trout. The whole area is beautiful and the river's a fisherman's dream.

The wild steelhead are a rare breed. These trout spend three to five years of their youth in fresh water, swim to the sea for a few years, and then return to the rivers and streams to spawn. They look similar to rainbow trout but have a more streamlined body that looks like a torpedo. When they've come from the ocean, they're bright and silvery and can be ten years or older and weigh over thirty pounds. When you find wild fish living in these rivers, they're an amazing sight. Their surroundings are a paradise because there aren't any dams and they haven't been changed through hatchery programs.

The BC government did the right thing in 1987 when they protected these fish. Before '87, you could take a maximum of eight fish per person from the river. If the government hadn't stepped in, the fish would probably be gone, but because of the regulations, people from all over the world can visit and find the king of fish still running wild. Since 1987, the sport fishing's been catch-and-release, so we've been able to keep this natural resource healthy: we catch them, revive them in the river and watch these special creatures swim off.

Joy Hinter, Marilyn's friend from Langley High School, introduced us to this remote wilderness in the mid-'50s. Joy had taken a job at Norlakes Lodge (later Babine Norlakes) cleaning out the cabins, washing dishes and doing laundry. She planned on working there just for the summer, but she met and married a good-looking Danish guy named Ejnar Madsen. Joy was eighteen and Ejnar was twelve years older, and they made a great pair. Like us, they married in 1956. Ejnar soon became a partner in the Norlakes fishing lodge and by the mid-'60s had built cabins at Steelhead Camp a few miles downriver.

In October of 1956, Ejnar and Joy invited us to stay at Norlakes. To get there, you need to go by boat or float plane. We were overwhelmed with the area's beauty, and the four of us became such great friends that Marilyn and I would travel to the Lodge every year. We'd help Ejnar and Joy close up camp in late October and then, when they were still living in Surrey during the winter months, he'd do extra jobs like working for me in the mill or building houses to sell later. He was a good handyman and carpenter.

The first time Marilyn and I visited, we quickly became familiar with the area and knew we'd be returning as often as possible. That first trip, the four of us walked the three miles from Ejnar and Joy's lodge to the Department of Fisheries' fish-counting weir, and then we carried on along a trail as far as the Gravel Bar Pool to where the Nilkitkwa River flows into the Babine, just north of Nilkitkwa Lake. In 1946, the Department of Fisheries and Oceans built the weir a mile downstream of the lake, and the fisheries set up a camp nearby to count adult sockeye salmon. After tagging the fish that swim up from the Pacific Ocean, they open the trap doors to release them into the lake. The Natives from the nearby Kisgegas Indian Reserve take their fish out of the weir, but we fished along the river with spinning rod and reel for steelhead. I remember the first October we were there: the weather was pretty cold, but the crackling fire back at camp kept us warm. The whole experience out in nature had such a good effect on us that we promised ourselves to take short vacations to the Babine as often as we could, especially when we knew Ejnar and Joy would be our companions.

The entire area is wild with bald eagles, moose and grizzly bears all living in the four-thousand-square-mile Babine watershed. Sometimes we'll see the bears come out of the forest to catch fish on the banks of the river. They want the steelhead and compete with the fly-fishers. They stroll along the shore with their cubs, swishing the water with their giant paws, throwing the fish out on to the beach for their babies. When they're in the water and you're fishing in the river, you can either get out or keep a safe distance.

The people fishing usually know when a bear's around because they can smell it. Most of the time, if I see a bear upstream I'll keep on fishing because if you don't bother them, they won't bother you, but if one looks to be getting too close I bang two rocks together to scare it off, and they usually leave. Other fishermen prefer to make a quiet exit and let the bears have the river to themselves. Luckily, the bears don't go to the water often. When they do wander in, they might ignore the steelhead and get hold of salmon that are dead after spawning.

At Ejnar's camp one time, Marilyn, her dad Mike and I were guests along with a few Americans. We were still allowed to keep the fish at that time, and the Americans wanted to take theirs home to get them mounted. They put their catch inside a large pit dug deep in the ground to keep the fish cool, then covered the pit with a steel lid. Unfortunately, a black bear had come by in the middle of the night, lifted the lid, and eaten their fish. In the morning, the men were disappointed and wondered what to do to keep out the bear. Ejnar was also upset that the bear was robbing the underground fish store. I remember him telling me, "We gotta get rid of that bear."

We decided that one of the guides and I would take watch for the first night. We stayed inside the cabin with the door opened a crack. I had my father-in-law's gun at the ready. The whole place was dark, but just after midnight we spotted the shape of a bear coming down the trail. The bear got to the pit and the guide opened the door wider, so I could take aim. I shot him, and the sound echoed like a big explosion. The bear went down fast. As we stepped toward it to make sure it was dead, the rest of the guests came running out of their cabins onto the snow still dressed in their pyjamas. No one had known we were going to shoot a bear, so they wondered what the heck had made that loud boom. As everyone returned to bed, Ejnar and I put it on the boat, motored several miles and dropped it over the side. The American guests were glad to keep their fish that year.

Another time we were sitting on the lodge porch having a beer, and we saw a mother bear with her three cubs on the opposite side of the river. The place was quiet and we kept still as we watched this wild family go to the river's edge and sniff around. One of the cubs was more adventurous, and he wandered farther from the others and got distracted, so he never noticed when they moved on. We could see him look around for them and become frightened. He hurried in their direction and soon caught up to his mother. She must have known he would show up because she hadn't looked worried.

The watershed is also home to moose. A common sound is a bull moose crashing through the bush a few yards away. We would sometimes hear one coming closer but we wouldn't panic because he wasn't interested in us—he'd come bounding out of the bush and plough across the river and carry on. Amazing to see.

One time, Ejnar had to go to Smithers to get a part for his boat's motor. Along the narrow logging road, he came up behind two moose hunters in their truck driving about ten miles per hour to avoid scaring any moose that might be up the road. Ejnar tried everything to get around them, but they wouldn't let him pass. Eventually, the road widened and Ejnar managed to get past. He sped ahead, leaving the truck crawling along. As he came around a corner, he saw a big bull moose right in front of him. Of course, he laid on his horn, and that was that for the hunters.

For several months after Vic and I became partners, I'd been telling him about the Babine. In 1964, we had saved up enough money to stay at the Norlakes steelhead camp as paying guests. We had been saving a dollar a load from the truck drivers who picked up our sawdust and used that money as our fishing fund. We drove up from Vancouver to Smithers and on toward Babine Lake.

When we were driving on one of the logging roads that led to Ejnar's camp we had decided to stop to test our rifles. We crossed a bridge over a small lake and took a couple of potshots at a tree when, suddenly, two hunters came racing out of the bush toward us.

The bigger one yelled first: "You're not allowed to shoot here. I'm gonna report you to the game warden. Tell him you were target shooting."

The smaller guy came along and added his voice. "You're shooting off the road. That's not allowed."

"Yeah," said the big guy. "You're scaring the moose!"

They had a camp set up nearby, but we hadn't seen it. Vic and I apologized and explained we were heading to the Babine to fish. We walked back to our truck, and they returned to their tent.

We carried on to Babine Lake, where Ejnar had left a boat at the dock for us, with a note saying, "Come on down yourselves." We motored down to the river camp, where we fished for about four days. On the way back in the boat, I had my rifle and we taxied near the side of the lake, where I saw a moose on the riverbank. I climbed out of the boat into the shallows and shot him. I always aim for the head or under the shoulder, so down he went. We sliced him open, gutted him, cleaned him as best we could, and put him in the boat. We lifted him into our trailer and covered most of him with a tarp that we tied down, but his hind legs were still hanging out.

We drove back the way we came and saw the two hunters near the road by their campsite. We slowed down alongside them so they could get a good look at the moose. The big guy was walking along the road with a rifle and didn't see it at first.

"Did you have any luck fishing?" he asked.

I said, "Did you tell the game warden about us shooting?"

"Nah." He was friendly until he noticed the legs of the moose sticking out from under the tarp. "You got a moose! Thought you were going fishing." He came closer to the trailer. "That might've been our moose if you hadn't chased him away with your target practice. You never said you were moose hunting."

"Had no choice," I said. "The crazy thing jumped in our boat." I hit the gas pedal and sped off.

Vic and I drove for another sixty miles until we got to Smithers. We decided to give the mill a call to see how things were going. Marilyn answered the phone and said, "Emil's in a panic. We didn't have any way to get in touch with you."

"What's wrong?"

"The motor on the head saw's broken. The mill's been down for two days!"

"Why didn't Emil buy another one?"

"He said you never authorized him to buy machinery."

I talked to Emil and told him what to do: "Buy a fifty-horsepower motor and get the mill going as soon as possible."

Marilyn was new at the game and unsure what to do. "We couldn't make any decision without you or Vic," she said, "and Emil didn't want to take responsibility for spending money."

When we got back to the mill, Vic and I reassured everyone that we'd never go away at the same time. Not until years later did we go on a trip together when we travelled to Japan. By then, we were better organized and had millwrights who could make those kinds of decisions without Vic or me at the mill. Vic and I returned to the Babine but separately after that incident.

In September of 1966, we hired Claud Muench, who we knew from Bob and Ian MacDonald's mill. Claud had been a millwright's helper and had been driving for Shell Oil out of Cloverdale for a few years, hauling fuel into S & R. He knew his stuff. Plus, he had his first-aid ticket. One day when Claud was in the yard, he stopped me and said, "When are you gonna give me a job?" He wanted to work on the boom or in maintenance.

I told him, "We need a millwright." S & R was indeed growing so we needed to hire workers.

"I have to give Shell a couple weeks' notice."

Claud was an invaluable employee at the mill, but he also helped me years later when Marilyn and I became more involved on the Babine.

Steelhead fishing on the Babine was a special sport in those early days before catch-and-release. The first change in the regulations came in 1970 when the fisheries banned the use of bait, like salmon or steelhead roe on the Babine. During most months, we might find one hundred salmon in the river but only six steelhead. The salmon come up from the ocean at Prince Rupert and continue along the river to spawn and die; the steelhead come up the river and return to the ocean. The brief window of opportunity to catch them comes in September and October. When we first started fishing the Babine, anglers could keep their fish and they might catch two to four in one day. Ejnar had built a large boat, almost like a houseboat, to bring twelve fishermen down to the main lodge. During one of the last years that the fisheries allowed us to keep the fish, Ejnar almost cried when he saw them take their quota of eight steelhead each. They had packed and frozen ninety-six fresh steelhead in one week. He was afraid if they kept this up they'd reduce the stock to zero.

Besides the limited number of fish, steelhead fishing is in another class of its own. We had to be good fishermen, and in the case of fly-fishing, a good practice was casting a long line with only one false cast. A false cast is when the fly line is kept in the air before dropping it in the water. We might need to false cast to change direction or lengthen the distance on the cast. Nobody wants to false cast several times behind them and have the hook catch on a tree. We also had to be in the right place in the water; the river is about twelve feet deep, so we stay near the shore, but we might need to walk half a mile to find a good spot.

Several years after our first trip and after Vic had retired, the two of us were guests at Ejnar's camp. Todd Stockner was guiding us, and we stopped the boat at a likely-looking spot on the way down to Callahan's pool. He let us off to have lunch and said, "Good luck. Nobody's ever caught any fish in there." The water was coming through a really fast lane and churning, so Todd figured it was a lost cause. While Vic finished his lunch, I put my line in the water and waited. Within twenty minutes, I'd hooked three steelhead. When Todd returned to pick us up, he put his line in and caught another one. The pool didn't have a name, so after that it became known as Chick's Chute.

The guides drop off fishermen at a number of spots along the river. Steelhead find a place they like and they stay there, feeding in

Vic holds another beautiful steelhead before releasing it back into the Babine river. This photo was taken in the late 1990s when he would have been close to 70 years old. His expression shows how much he loved this sport.

these pools. Sometimes we can't see the fish, but we throw out the line and the fish bites and all hell breaks loose. The other guys have to take their lines out of the water to avoid getting tangled while the fish fights the line.

Vic became well known on the Babine because he was a good steelhead fisherman. He had started out using gear but when he took up fly-fishing, he never went back. He caught hundreds of steelhead and had a graceful casting technique. Some guys might not catch any; they'd get skunked. They could hook it, but as they brought it in the fish jumped the hook. Vic was successful every season.

People also knew of Vic because he used the Bubble Head fly. Vic first read about Bubble Head flies in *Salmon Trout Steelheader*

The Babine Steelhead Lodge after construction.

magazine in Oregon. He tied a few up and tried them out in 1982. The Bubble Head attracts the fish both because of its colour and the wake that trails behind the fly. The Bubble Head has a fan shape of deer hair left in front to form the head.

I used a red corky that floats. I'd throw it upstream and watch it on top of the water. Then a fish would grab it. I loved the feel of the tug on the line, and I'd have to be pretty fast to hook them. The water would be flowing one way, and the fish was going in the opposite direction, so I'd give it a good snag to set the hook. It could take twenty minutes to bring in the fish. Fly-fishing means that you tie the fly yourself. Sometimes at the mill office, I'd walk by Vic's open door and see him busy tying flies. He felt good when his handmade fly caught the fish. I never used flies; I could catch twenty a day with my corky. Fly-fishing is more of a skill.

When the regulations stipulated that we could only catch and release the fish, we had to learn the knack of keeping the fish alive. This required you to pull it to shore, lift its head quickly out of the water, remove the hook, maybe take a photo, and then let its body slide off your hands as you release it back into the river. Vic had been using barbless hooks and talked to the fisheries people to encourage them to enforce barbless hooks. A fish gets stressed when you take the hook out with pliers. With the barbless, you just squeeze it and it comes out right away.

After years of staying at Ejnar and Joy's lodge and enjoying the fishing, Marilyn and I had an opportunity in 1987 to buy our own place about six miles downriver from their Norlakes property. Bob and Jerri-Lou Wickwire, both well known in the community, had built and owned the lodge for many years before selling it to two American doctors who wanted to sell it after only a couple of years. We paid them cash and told the fisheries in Smithers that we had bought the fishing lodge and were going to call it the Babine Steelhead Lodge. Fishing season is only during September to late October, so Marilyn and I figured we could take on the extra work of owning our own lodge while still running the mill.

In 1987, we had only six months to prepare for the season. I called on a number of guys from the mill to come help. Claud Muench, who was now a foreman at the mill, flew up to Smithers with me and helped me get the place in order. We also hired the Spence brothers' company, called R4, to help us get the camp into shape. The river was eroding the bank, so we needed to build it up. We had a truck full of sand parked at the weir, and we filled nine bags with sand and hired a helicopter to carry them in a cargo net back to the cabin. We loaded and unloaded about ten times before the bank was four or five feet higher. We then trucked in two forty-foot-long trailers of building materials and equipment from the coast and had all that flown to the lodge by helicopter. We raised the cabins and the main building on skids and moved them back from the edge of the river. We brought in the cookstove by boat and had to take the door frame off the main lodge to get it inside. We also added a dining room, drying room, washroom and indoor toilets. Originally, the cabins had pot-bellied stoves, but the place would get messy with guys bringing in wood and not knowing how to keep a fire going, so we put in gas heaters.

We had a small 5,500-watt generator in those days, and a couple of years later we bought a much bigger 7,500-watt generator the size of a farm tractor and kept it in its own shed behind the cabins. The original generator became our backup. We were deep in the bush and relied on crews from Smithers to keep the lodge functioning during the cold and dark.

We knew we were in the middle of a wilderness each time we saw a moose or a bear. We used to dump the garbage behind the lodge in a big pile to burn it. One day, Claud was walking up the trail behind the guest cabins to go to the guide's cabin when he spotted a grizzly bear several yards in front of him. A sixteen-foot bank was behind the bear. Claud was about forty feet from the safety of the cabin and was wondering if he could run fast enough to get there before the bear had time to catch him. The grizzly was looking at him and, even though he knew not to look a bear right in the eye, he couldn't take his eyes off him. Suddenly, the grizzly turned to the bank and in two jumps made it to the top. Glancing back at Claud, the bear picked up a two-by-six board and hurled it through the air like a helicopter to show who was boss. The bear turned and ran like a freight train through the bush. Claud stayed frozen for a minute as he realized the bear had given him a break.

A view of the Babine river from the Babine Steelhead Lodge shows the jet boats tied up for the night.

We learned that if a grizzly was around, the black bears would be gone. The grizzly puts a run on the black because he's a superior animal. Luckily, no one else had that kind of encounter with bears.

The steelhead arrive in the river in the summer, but the fishing season starts mid-September. The season carries on until the last week of October, and the weather can get pretty cold.

In the spring, the steelhead stay in the river and lake to spawn, and the season's process begins again. Marilyn and I were set to open the lodge but had no idea that we were starting at the worst possible time.

11

BABINE
STEELHEAD LODGE

ARILYN AND I were fully involved in the whole operation at our Babine camp from September to early November. We hired a cook and his wife and got three guys from our mill—Dan Steiner, Brad Siemens and Dean Schroeder, all good steelhead fishermen—to take time off from the mill to work on the Babine for the same wages. Dean Schroeder still works at the mill.

We had only one mishap later on with a new guide. In the beginning, we were using plywood boats that don't get stuck on the rocks (the newer aluminum ones can get stuck). It's tricky running the river and missing the rocks because we have to keep between them. This one time, the new guide hit the rocks and the boat took in a bunch of water. We had to tow it to shore and get it pumped. Other than that one sinking, our guides always worked well in the boats.

The first and last weeks of the season, Marilyn and I were there before the guests to open and close the camp. We had learned what to do from watching Ejnar and Joy. The first week, we'd take into the camp eighteen to twenty 45-gallon drums of gas for the boats and ten to twelve propane tanks to heat the three cabins. Each cabin had four bunks to accommodate our twelve guests and two propane tanks for heat. Our guests only stayed one week, and we had a pretty good system to get them in and out of our lodge.

Over the length of the season, when guests were arriving we would take the weekend off from S & R and fly to Smithers on Thursday night. We'd drive five miles from the airport and stay the night in a motel in town and then phone the camp guide or the cook for the food list. We'd do the shopping on Friday, which usually meant getting jugs of water, ham, steaks, chicken, vegetables, milk, cereal, bread and eggs. Marilyn and I would take two separate vans and drop off the food at the weir and pick up the guests who were coming out of the camp. Saturday morning we'd pick up the new guests at the Smithers airport. We'd drive the eighty miles from the airport to the main road and on to the river where our boats were moored, and motor to the lodge. The journey from Smithers to the camp took about three hours. On Sunday Marilyn and I drove the two vans back and parked at the airport and flew home. Our schedule was hectic, but we always had time to fish.

The drive was slow on the logging roads because we had to watch for oncoming logging trucks. The sawmills cutting logs from the area were in Smithers and the roads weren't usually that busy, but when a truck occasionally did come booming by, we needed to be alert and get out of the way as best as we could. The drivers are paid by the load so they're always in a hurry. One time, we got to the top of a hill and were about to take the corner when we saw a truck coming toward us. We tried to hug the other side and got nervous when one of the logs sticking out at the back almost hit our van. I could have rolled down the window and touched it; that was uncomfortably close. Sometimes, while on the river, we'd see ten to twelve trucks going over the bridge across the upper Babine while the fish were running. Fishing and logging were good for Smithers, as was the tourism during fishing season.

Two beauties! Marilyn holds a steelhead long enough for a quick photo. Moments later it's back in the water.

When the guests arrived at the camp, they'd unpack and come into the main lodge. We had advised them to bring chest waders, three or four different rods, tackle, lines, single-hook lures and warm clothing, which included long johns, wool pants, heavy wool socks, shirts, toques, gloves and fishing vests, as well as good-quality rain clothes, camp shoes for the evenings, insect repellent, personal toiletries and their own liquor and mix.

The guests would get set up with the three fishing guides, who would each take four of them to different parts of the river to fish. The guides communicated with each other through their radios. One would ask, "How you doing up there? Any fish?" The other guys would answer, and if they had good spots, the guides would move the guests around to avoid overcrowding in popular pools and to make sure everybody had a chance for a successful day. We wanted everyone to get a chance at each run to ensure they caught some fish.

Our routine at the lodge was pretty simple. Claud had set up a timer system so when the cook got up at 5:00 in the morning, he would start the generator but not turn on the lights; they'd come on automatically about an hour and a half later. We'd get up around 6:30 or 7:00 and have a shower. (We had separate showers for men and women.) Most of our guests were men, but once in a while we'd have women who had come out to fish. In the main lodge we had a pot-bellied stove, so I always checked it first thing to make sure the fire was lit. I liked to chop wood, so during the day I'd be outside adding to the woodpile, making sure we had enough. Late September, the area started to get a little cool. We had two sofas, a few chairs, a TV, and a satellite radio in the lodge. Our guests wanted the place to remain rugged. They'd come in for breakfast and rub their hands over the stove before sitting down for their meal. When they had finished eating a hearty breakfast, the guides would take them to good fishing spots.

Our guests never interfered with the local fishermen from Smithers who fished the lake for trout. The locals who didn't own boats would drive to the weir and park and then fish all day with rod and reel. Everyone had a fishing licence and could keep a few of the trout.

After a day on the water, our guests would congregate in the lodge in front of the fire and drink their beer or wine and swap interesting stories. They'd also discuss how many fish they caught that day. If they had caught a numbered fish and released it they'd be excited to recognize the number if they caught it again: "I remember catching that same fish last year." I always enjoyed seeing a tagged steelhead in the water too. I'd look and say, "Here comes number thirty-five again. I caught and released him last year." Fortunately, these beautiful fish don't die in one season; they return year after year.

In the evening after dinner, the guests would get their gear ready for the next day. At night, the generator and the lights would automatically go out at 10:00. The guests would wake up at 6:30, put on their duds, eat and be out on the river by 8:00 and back to the camp by 5:00 p.m. If they wanted to continue fishing, they could fish in the home run outside the cabin while their friends sat on the porch enjoying a drink.

On Sundays, Marilyn and I would head back home. We would take the motorboat from our camp to the parking lot, drive to Smithers, park at the airport, fly back to Vancouver, sleep and wake up in time to be at the mill Monday morning, where we worked until Thursday and did the whole journey up to the lodge all over again. We had a pretty good system.

Our guests came from all parts of the world. We had Americans, Europeans, Canadians, Japanese, both men and women—and they all loved the area. Four Japanese men were regulars and came on the same week each year. They owned a McDonald's franchise and spoke pretty good English, so Marilyn and I got talking with them during some of their discussions. We had ads in fishing magazines, but people also heard about us through word of mouth.

One year, a regular guest from San Francisco came up for a week and on this particular trip bought a cap with our logo. He had been fishing at our camp with a group of eleven other guys every year for some time. A couple of months after buying the cap, he was riding a trolley car with his daughter and she was wearing it. While they sat on the tram, another guy standing beside them noticed the Steelhead logo and started up a conversation, and they got talking about fishing. The man with the daughter pointed to the cap and said, "I got that at a lodge on the Babine. I fish at Chick Stewart's camp."

The other guy said, "You're kidding me. I know Chick."

"From fishing?"

We catch these beautiful steelhead before releasing them back into the Babine.

"No. From the mill. In fact, I'm going to Vancouver tomorrow to see him."

Turned out the guy was a lumberman and he was coming up to see me about lumber scows. He got a kick out of telling me he'd met one of my guests on a trolley car.

Those weeks greeting our guests and working through the season went fast for Marilyn and me. We had the energy to keep it up, and it was always fun except for the very first year we had guests.

The previous owners had already booked our season's guests, but the fishermen still had to pay for the week. In that first group of twelve men, we had four doctors from Chilliwack and a party of eight from Montana. Unfortunately, this just happened to be the year the BC government brought in the catch-and-release regulations that stipulated fishermen could no longer keep their steelhead. Marilyn and I were worried that we'd just bought this beautiful lodge and we'd never get any customers. Who would want to pay

This picture of Marilyn was taken on our boat near the Babine Steelhead Camp in September 1990.

to fish and then return their catch to the water? When we told the guests about the change in regulations, they were far from happy because each of them had expected to take home trophy-sized Babine steelhead.

The tension around the dinner table that night gave me indigestion. Marilyn was almost in tears. The four doctors from Chilliwack sided with us and understood that the law was the law, but the Americans were not buying it. I pushed my chair back from the table and said to the eight of them, "If you don't pay me by seven o'clock tonight, I'll take you all back to Smithers first thing tomorrow morning."

"We don't see why we can't take a couple out of here."

"Yeah, who's going to see us?"

"I know this isn't great news, but my wife and I have just started the business. We don't want to lose our fishing licence on day one."

The Canadians paid up, but the Americans kept stewing.

"My wife and I will be in our cabin," I said. "You've got 'til seven to pay us."

We sat and waited. Marilyn and I were concerned that we'd gone to all this expense and effort and the whole operation might have to shut down. I tried to reassure her but I felt like crying too. At five minutes to seven, one of the Americans came to our cabin door and knocked; when I opened the door, he threw cheques and loose bills into the air. "We're staying."

Our first hurdle was crossed.

The next problem occurred a few years later in early winter when I got a call from Smithers at the mill office. Bob Hooton was calling me. Bob was a knowledgeable steelhead biologist overseeing fisheries management throughout BC, and between 1986 and 1999 he was the senior fisheries guy for the Skeena Region under the Ministry of Environment. He was also a strong supporter in the fight to protect wild steelhead. We'd sometimes see him tenting along the river with a friend, and they'd be fishing together. One time we invited him for dinner at our camp. Even if he was off duty, he could kick you off the river if you shot a moose out of season. Bob was committed to doing a good job and, on this particular day, he was calling to tell me he had been motoring along the river when he saw that a tree had fallen on one of our cabins and ripped a hole in the roof. Lucky he was on the river because the camp was closed for the season.

"If you don't fix that hole, the snow'll get in and it'll be an absolute mess inside."

I thanked him for letting me know and got into action. I asked Claud to fly to Smithers with me, where we rented a helicopter to take us to the cabin. The helicopter was able to land at the main cabin, so we got two power saws, a can of fuel, and hip-waders before heading to the damaged roof. When we got to the small cabin, the pilot couldn't find a place to land, so we dropped off the saws and slid out and jumped into the river by the edge of the water. I told the pilot to come back for us in two hours. He found a clearing a ways downriver to wait.

Meanwhile, Claud and I got up on the roof and started cutting the tree into pieces.

We rolled the logs onto the ground, brushed the snow off the roof, and found tin sheets to patch the hole. We locked the saws in the woodshed because we didn't have time to go back to the main lodge. The helicopter returned right on time and hovered above the ground. We climbed in, flew back to Smithers, and caught the plane back to Vancouver, all in the same day.

We didn't have too many emergencies like that, and not too much could go wrong when we fished steelhead. It's about as safe as fishing salmon, except we weren't fishing out in a boat; we'd stand in the river in our hip waders and keep the rod in our hand at all times. Very little action until we felt a bump on the line. One time, however, I was getting out of a boat to fish at Corner Pool, and I wasn't paying proper attention. Instead of stepping on what I thought was a rock, I stepped on a dead Chinook salmon and slipped and fell on my back in the river. Soaked right through. Fortunately, I didn't hurt my back.

In fact, the only time I got hurt on the Babine was one day when I went up to look at the big cabin that was above a set of grated steps. One of the steps was off a bit, and I got my foot caught, which sent me flat on my face. I had a huge gash in my forehead that was bleeding quite a bit. Luckily, our chef at the time had taken a first-aid course, so he tried to fix it the best he could. My daughter Wendy

TOP This photo taken from the helicopter was our first glimpse of the fallen tree lying across our cabin's roof.

BOTTOM What a mess! We were lucky the tree hadn't completely destroyed the cabin.

drove me into Smithers to get stitched up, and I remember her going over the gravel road a lot faster than normal. If we ever did have a serious medical emergency on the Babine, we were prepared with a landing pad nearby for the medevac helicopter.

Marilyn and I also liked to fish steelhead in Washington State. We'd often travel about 240 miles to Forks, Washington, and fish for the weekend. Our girls were grown up, so we'd head down there on our own for a break from the mill. We got the ferry and then drove the rest of the way. We had a guide on the river who'd take us in the boat to a good spot to fish. When we went through the border one time, the lady at the crossing asked us where we'd been, and when we said "Forks," she asked, "How many fish you catch?" I answered, "Six." She nodded, "My husband's out fishing. I'd be impressed if he brings that many home."

Another time, I guess Marilyn and I had been working pretty hard because when we were settled on the ferry, we fell asleep. We stretched out on the booth seats of a table for a quick nap and fell into a deep sleep, so we never heard the call to return to our car. A crew member had looked through the restaurant but couldn't see us, so he got the other ferry passengers to drive around our car. We finally woke up and saw we were alone. We raced downstairs. When we arrived on the car deck, the place was empty except for our car and the workers, who applauded our arrival.

One time we were heading to Forks, but first had to go to a meeting in New Westminster. We were in our work clothes, and when the meeting finished we figured we'd better stop at the bank to get some cash for the weekend. We went to a branch near the meeting place and I made out a cheque for $100. The teller looked at the cheque, looked at me in my grubby work clothes and said, "I'll be back in a minute." She went to her desk and picked up the phone and dialled. She spoke, nodded and hung up the phone, then came back to her till and handed me $100. I asked her who she had to phone to get permission.

"I called the manager at your bank and asked him if I should cash your cheque."

"What'd he say?"

"He said if it's not more than a million dollars, cash it. Mr. Stewart's good for it."

During our wonderful trips to the Babine before we bought our camp, we always left with great memories of our stay with Ejnar and Joy. In the early 1980s, however, we received bad news: Ejnar was diagnosed with cancer. We were all shocked when he got sick, and we kept in contact when he came to Vancouver for his treatments at the Vancouver General Hospital cancer centre. I would go up and visit him or take him to the airport to return home. In 1983, when he returned to Smithers for the last time, we took him out fishing; he wanted to be on the lake because it had been such a big part of his life. But for the majority of the day, he slept in the boat. He said he didn't want to leave his life so soon, but "it was the way it was to be." He died on September 13, 1983. He had a great sense of humour and real love for his family and the outdoors. Marilyn and I continued to miss him, and I still miss him. It was rough for all of us who knew him to think of him going so soon. Ejnar had asked that he be cremated and that his ashes be scattered in the Lower Forty Pool on the river.

Joy and their eldest son Karl ran the lodge for another three years with the help of their other children, Erik and Karen, before selling it in 1986 to Christopher (Kit) and Hazel Clegg, who had been coming to fish the Babine for many years. They had told Ejnar that if he ever decided to sell, to speak to them first. Their son Pierce and his wife Anita from Smithers later owned and guided the Norlakes Trout Lodge and Steelhead Camp.

A few years later, we all became involved in an issue that threatened the way of life on the river. Once a month, a government fisheries guy would come by to make sure we were abiding by all the regulations. We never had any infractions, but we did have a political issue that got all the owners of the lodges together to find a solution. Pierce Clegg was always a strong supporter of the river and its fish, so when he found out that the forest ministry was planning to build another logging bridge called the upper bridge near a perfectly good bridge that already existed, he started asking questions. The government had approved a twenty-year plan that would increase clear-cutting in the area, and they needed access roads and more crossings for the logging trucks.

I knew the timber guys, and my own business depended on logs being cut, but the Babine fishing resource not only provided millions of tourism dollars, it represented a natural wilderness that needed our protection.

At the end of our first fishing season in 1987, Pierce and I travelled to Victoria to speak with politicians and explain how bad this logging bridge would be for the area. We met with the ministers of environment, tourism, recreation and culture, the deputy minister of forests and lands, and the assistant deputy minister of operations for the Ministry of Forests and Lands. The logging company had said that they needed to build another bridge because the existing bridge and road were in a terrible state and would be too expensive to repair, so the better plan was to build new crossings. This was not true: the bridge was in good shape then and in 2017 is still in good shape. Plus, the forest service had spread a rumour that the pine beetle had taken over the specific area the logging company wanted to log, but local forest companies admitted this rumour was false, and the pine-beetle threat was minor. We explained all this to the politicians; they agreed to set up a committee to look at the issue and assured us that, while they investigated the logging company's proposal, they would hold off building any bridges.

Months went by so, to add some strength to our fight, I contacted Bill Vander Zalm, the premier of BC at that time. The premier's political career had started back in the 1960s when he was alderman and then mayor of Surrey. Our paths had crossed over the years at community functions and when Marilyn and I would buy our flowers from his Fantasy Gardens nursery. We always got on and had friendly conversations, so I decided to call him about the planned crossing. He listened while I explained that the second bridge would kill the fishing and the existing bridge was only a mile away from where they wanted to build another one. I gave him the names of the ministers Pierce and I had met in our meeting, and Vander Zalm told me to leave it with him, that he would look into it. More time passed before I received a phone call from him saying the bridge would not go through on the upper Babine.

Pierce received confirmation of this from the Ministry of Forests and the Ministry of Environment in September of 1988. The press release said the plans for the bridge had been cancelled because a study showed that the Babine River had special fishery value.

We were all glad to hear the area would stay the way we'd known it for decades. When I took time away from S & R, I was happy to fish and hunt in such a beautiful place—but nature can also be dangerous. One fishing trip on the Babine reminded me our neighbours are often wild animals.

12

CHILLS AND THRILLS

NE FALL WEEKEND, I invited our friend Jeff Hobbs and his son Bill to join our other guests and come fish on the Babine. Bill was in his early twenties, training to be a realtor like his dad and working hard, so a break from the office was just what the doctor ordered. In the cold mornings, the river is sometimes covered in mist and the trees are red and yellow, so the whole place looks peaceful until a wind blows in, the leaves start swaying back and forth and the river roars outside our cabin. The cold and the damp didn't slow us down though; Jeff and I could hardly wait to step into the currents and start fishing.

We left camp in the morning to find a good fishing pool. We motored along in one of our skiffs, which is like an oversized canoe about twenty feet long with a 50-horsepower jet motor. We used these boats to deliver fishermen to the fishing grounds. I found a good-looking run, so we pulled the boat up to shore and threw out the anchor.

Jeff and I started fishing downstream, but Bill decided to stay by the boat and catch a little shut-eye. He settled in a comfortable spot leaning against a rock and fell asleep. Jeff had just sent out his lure and plunked it down in the river when he hooked a steelhead. The powerful fish was thrashing around, fighting to escape, but Jeff kept playing it; I hurried over to help bring it in.

I can't say what made me look upstream, but it might have been an unexpected movement along the riverbank caught my eye. I strained to make sure I was seeing what I thought I saw: about ten feet from sleeping Bill was a massive black bear sauntering toward our skiff. I had to decide—do I stay and help Jeff bring in a beauty of a fish or do I leave him and help Bill? It was either the fish or the son. As a fisherman, I had a hard time making up my mind, but as a dad I figured I'd better go help the kid. I needed to wake Bill and warn him the bear was there.

As I ran toward the skiff, I noticed a bear cub several yards away. The adult bear was obviously a mother and wouldn't take as long as I had to decide who to attack first—Bill was her closest threat. I knew not to surprise the bear, so I slowed down, picked up a few rocks and started banging them to let her know I was approaching. Bears have an excellent sense of hearing. They also have an excellent sense of smell. The animal was now inside the boat, sniffing Bill's lunch kit. The same moment that the bear saw me, Bill woke and jumped to his feet. "Don't eat me! Don't eat me!" he cried as he slowly backed away. Fortunately, he was moving away from where the cub was waiting. Bears can run as fast as a horse and are obviously strong; they'll also defend their area, their food and their young. The bear turned her head toward the cub. I wasn't sure what I would do if she decided to charge, so I kept back and waited. She must have thought it a good idea to avoid problems with the humans because, instead of being aggressive, she stepped out of the boat and casually walked off into the bush with her cub.

Jeff had lost the fish and left the fishing hole to check on his son. Bill was pretty shaken up, but amazed the bear had just wandered away. When his heart rate got back to normal, he thought the whole

My pack horse is loaded for a week of camping and hunting in the back-country near Smithers.

thing was quite an adventure. We looked inside the skiff and saw her big muddy paw prints in the boat.

Jeff said, "Won't take a minute to clean that up."

I had other plans. "Leave it. If we go back to the lodge with this story, they might not believe us. Those big paw prints are our proof."

Sure enough, the guys at the lodge weren't convinced the bear was as big as we said, so we marched them down to the skiff and showed them we weren't exaggerating. They were impressed. We were all glad nobody got hurt and reminded each other to stay alert. We had a few good laughs that night as Bill explained that his best defence against a mother bear was to back away and speak English: "Don't eat me! Don't eat me!"

For years before we bought our own lodge, the territory around the Babine had drawn me back every fall. If I wasn't fishing on the river, I'd go up to Smithers and hunt moose in the mountains. I kept this up most years after Marilyn and I had bought the lodge and closed our camp for the winter. We had four boats: one to take us out of the camp to the parking lot and three for the guides and guests. We'd remove the motors and leave the three boats upside down at the camp when we closed things down. Art Johnson, a Smithers guy, hauled fuel and had known us for years. He fished the Babine with us as well. During the off-season, he'd let us store our motor at his place and keep our fourth boat in his yard. Art owned pack and riding horses and had a guide's licence for taking out a hunter so, after we'd closed the camp, he and I would load up and go off into the hills in November to hunt.

We would trailer the four horses and drive to the base of the mountain near Smithers. After parking our trucks, we'd pack two of the horses and ride off to set up a base camp for a week of tracking and hunting. At the camp we unloaded and then tethered the pack horses before heading out with our rifles to find a moose or a big mountain goat. We scanned the mountains for goats and travelled the valleys for moose.

One of the things I always liked about hunting was that out in the woods I could live rough. The weeklong adventure toughened me up a little and made me think about the old-timers who slept outdoors, lived off the land, rode horses and got all that fresh air. I also enjoyed the meals cooked over an open campfire; the food smelled and tasted better. When we packed up the horses, we'd make sure we had the right rigging before hitting the trails; that way we were prepared and remained alert for any danger. The whole experience of tracking an animal was exciting and kept me sharp, with my .303 close beside me.

We took precautions at night to keep out unwelcome animals. When we unpacked the horses and gave them their food, we put a milk can under each horse's neck for its feed and tied a metal ball to the bridle. The tinkling noise kept away bears and wolves while the horses ate. The other predator to watch out for was the wolverine. They're dynamite! They look like a small bear but they're more like a weasel, and they're not afraid of anything—they're strong and have powerful jaws that can break bones, and their long razor claws will gut prey, so I never wanted to cross their path. They'll eat anything, including squirrels, sheep, deer and even moose. No one wants to take on a wolverine, but wolves will attack and kill one without too much trouble.

My first encounter with a wolverine happened during one of my trips with Art. He had shot a mountain goat and killed it. We were on a steep slope and he needed to slide down to get the goat, so he handed me his gun and gear. I was working my way around a small rock bluff with my arms loaded with Art's stuff, which included his gun and camera, and my own gun and binoculars when the wolverine suddenly came up at the top of the rock. It was a couple of feet above my head. It looked at me with its beady eyes and its mouth partially open, so I could see its sharp teeth; its long claws looked like they'd been extended just for me. I knew never to corner one of these animals because I'd have a fight on my hands: they'll attack until one of us is dead. So I stood there, armed but not dangerous. My arms were full of guns, but I knew I wouldn't have time to shoot. Crazy to admit, but my next thought was, "Please, don't eat me!"

Fortunately, Art made so much noise dragging the goat toward us that the wolverine backed off and disappeared. The other time I saw one was from a distance as it raced away into the mountains.

I managed to kill this moose on the second shot.

I count myself lucky that nothing worse ever happened between a wolverine and me.

Art and I could have just gone into this beautiful countryside for a trail ride and not hunt, but we liked to make it a hunting trip. During bull-moose season no one is allowed to shoot a cow moose, so we always kept on the lookout for the bull. As soon as I saw one in a clearing, I would sneak up quietly and get two shots away. The first shot would stun him and he wouldn't run right away, so I'd shoot again and he'd drop to the ground. We'd gut the corpse and straddle it over the pack horse and head back to camp.

The Babine and the area around Smithers gave me years of great fishing and hunting, but I also had some luck on the coast. In late spring of 1991, Marilyn and I signed up for the first Haida Gwaii Fishing Derby at Langara Fishing Lodge in the Queen Charlottes, where we planned to spend three days fishing spring salmon. We used to fish there even before the tournament started this annual tradition, so a contest sounded like fun. We paid an entry fee and joined about sixty other people in the hotel on the mainland before flying out to the fishing area.

Marilyn and I were on the same boat with a guide. He would hook up the lines, run the boat and help us fish. We had to wait until the organizer of the derby fired the starting pistol before we could begin. All the people who entered were to go in different directions on the water but stay within bounds, fish all day and eat lunch on the boat. Within a few hours, we saw two guys rushing in from their boat to get their catch weighed on the dock. They had a big one. We had barely finished talking about their fish when I suddenly hooked a salmon.

The guide wasn't allowed to touch it; I had to do all the work. I started feeding it the line carefully—if I pulled the line too hard it would snap. I held on tight and played the fish for about an hour and a half. I would've loved to reel it in but I couldn't—I told myself to just feed it the line and baby it along. When it stopped, I'd pull up the slack, keeping a tight line on it the whole time. If there was slack, the fish could take a run and bust the line, so I kept good control of the line. We had to follow the fish until it tired, so if it pulled us along we kept up with it. The guide was running the boat and Marilyn was watching to see if the fish was getting worn out.

Finally, the fish got tired. I brought it alongside the boat by reeling it in slowly. Marilyn grabbed the net and scooped it down into the water, surrounding the fish with the net so it couldn't escape. When we had it fairly secure, the guide helped us bring it into the boat. It flopped around on the floor of the boat a couple of times before it died. We quickly put a wet sack overtop of it. We rushed to register it at the main dock, where we found out it weighed fifty-one pounds, fifteen ounces. A pretty big catch. Marilyn and I continued fishing for the rest of the derby, but we never caught another fish that big.

This photo of me with my first-prize $50,000 catch was used in the Haida Gwaii Fishing Derby brochure to market the event for the following year.

At the end of the three days, around 2:00 in the afternoon, a gun went off to end the tournament. All the fish were in and had been weighed. My catch was at the top of the board, so every one knew mine was the winner. Second place was the fish caught just before mine, and it weighed three pounds less. Those two guys won $15,000. My prize was $50,000, so when we flew back to the hotel on the mainland, I bought drinks on the house at dinner. I gave the guides $5,000 and contributed another $5,000 to the Queen Charlotte Islands' Hospital fund. Marilyn got the balance of the winnings for our kitchen renovation.

Our fishing and hunting trips were a big part of our lives, but after several years of managing the Babine lodge, Marilyn and I decided to let our daughter Wendy and her husband Barry run the show. Barry had worked at the mill for twenty-five years before they moved to Kamloops, and they were happy to take on the responsibilities at the camp. Wendy booked guests and did the correspondence and Barry worked as a guide. The work involved twelve-hour days, but they were willing to take it over for us.

During the season, we helped them with the opening and closing of the camp, but they ran things throughout the eight weeks. Marilyn still took time to wander back into the woods to pick cranberries. She'd take her whistle to scare away the bears and a tin can to hold the berries. She had to be careful because grizzlies and blacks were often lurking on the riverbanks and trails. When the can was full, she'd hurry back to the kitchen and make cranberry jelly.

Barry and Wendy once had a scare over a grizzly bear. Barry was outside the main lodge talking to a new group of guests, and Wendy was at her desk doing some paperwork when their retriever Elmo scratched at the door to be let out. Wendy thought he wanted to be outside with Barry, but the dog ran past the back sheds and buildings and then let out a loud yelp. They ran toward the noise and saw a blond grizzly with three blond cubs. The dog had smelled trouble and wanted to protect his territory. Barry shot off his bear banger a few times, and the bear finally took off with her cubs. A few minutes later, they saw that Elmo had been wounded: the bear had taken a swipe at his backside. They got on the radio phone to Smithers to tell the vet they'd be coming into town with the dog. Everyone was pretty shaken up by the scare, especially the new guests, but Elmo survived and the guests had a memorable week on the Babine.

When Wendy first worked at the lodge she thought she'd feel isolated in a tiny cabin for several weeks, but the place is so beautiful in fall and the guests such great company that by the end of her first season, she was sad to leave. "I longed for the next September to roll around so I could be back at that special place. Even though we only saw the guests once a year for a week, they became our friends. I will treasure the memories we made with them." Wendy and Barry kept up the camp for the final years we owned the lodge.

We had been a part of the Babine for decades, but in 2016 I sold the lodge to one of our American guests. His partner from Smithers is now managing it. If Marilyn's friend Joy had never taken the risk to travel to such an isolated spot, she would never have met Ejnar, and we would never have discovered one of the most beautiful parts of our province. Ejnar and Joy were pioneers when they introduced us to the rustic scenery and great steelhead fishing. We had fun preparing the cabins for our guests—getting the fuel, checking the boat motors, chopping the wood for the campfires, adding the big generator, buying aluminum boats and meeting some fantastic people—but now it was time to call it a day. Another pioneering family, Pierce and Anita Clegg, have sold their lodge to Billy and Carrie Labonte, also from Smithers. The new owners will become stewards of this wild watershed and will continue to keep the steelhead fishing a top-rated enterprise for many years.

The fight over building a second bridge resulted in the creation of the Babine River Foundation (babineriverfoundation.com) and the Babine Watershed Monitoring Trust (babinetrust.ca), foundations that keep an eye on land management in the watershed and let the government know if any person or company is threatening the area. I'd figured my involvement in the political issues around keeping the Babine safe from unnecessary development would be the last time I'd have to work with politicians, but this was not to be. The next time I had to deal with them came when Marilyn and I were making plans to build a golf course.

13

THE KING OF GOLF

THE FIRST TIME I held a golf club was in 1953 when I was working at McDonald Cedar. The hockey player Babe Pratt was also an employee and he liked to golf, so on Wednesday nights in the summer, four of us would go over to Newlands golf course in Langley. Our foursome included Ted Smaback, a good mill man who'd lived in the area all his life, and the fourth guy was usually one of our customers. We would tee up and play eighteen holes, then go to the clubhouse for a few beers. Marilyn and I used to golf at Newlands because the course was a good place for average players. In fact, we liked it so much we had our wedding reception in their dining room.

I once got a hole-in-one on a par-three hole, which was pretty exciting, but I couldn't tell you how I did it. I gradually caught on to the game enough to have fun, but was never that good a golfer. I enjoyed playing with my friends and always tried to improve my game. My second hole-in-one came many years later when my grandson Josh and I were about to tee off at a course in Kamloops. Four other guys were waiting for us, so I hurried when I hit the ball. Josh yelled, "It went in! It went in!" The four guys clapped and I was a bit stunned. I think they took my quiet response as me being used to making those shots. Not at all! Never hit another hole-in-one again.

Golf became an important part of our lives many years after we played at Newlands. I guess you could say the road to that greater involvement started back in 1974 when our daughters were young and Marilyn and I decided to sell our house in Coquitlam. We wanted to find a home closer to the sawmill because we were getting tired of driving the long distance back and forth over the Pattullo Bridge, so we thought a home in Langley or Cloverdale would suit us better. I phoned a real estate agent and told him we'd like a house with a few acres, so he took us around to White Rock, Langley and then Cloverdale, where we saw a rancher-style house that included a twenty-acre farm.

I had always believed that property was a good investment because they can't make new land. After we moved from Tahsis, I got a tip from a guy that two acres had come available on the highway in Langley, a piece of property that I thought might work as a little retail lumber yard, so I bought it. I got another tip that six acres had come up for sale in Aldergrove and bought it as an investment. I figured no matter what I chose to do, the properties would come up in value. A couple of years later I sold the Aldergrove land and made a small profit.

Our new home would be more than just an investment: we wanted a nice place for the girls to grow up and extra land for me if I wanted to do a little farming. The owner of the rancher in Cloverdale had harness racing horses that he'd stabled on his twenty acres, but he wanted to downsize. He had sold the horses but the stables were still there. Marilyn thought twenty acres was a bit much to chew off, but I wanted to run some cattle, so we told the realtor we'd like to take a look at the property. The realtor would meet us at the house, and we would get to talk directly with the owner and ask him any questions we might have about the place.

The day was one of those dreary, rainy days we get on the coast. I had come from the mill and was wearing an old raincoat and a rain hat that were soaked from the downpour. I looked like something the cat dragged in, so I told the realtor and the owner that Marilyn could look inside the house, and I'd stay outside and take a wander around. I must have looked pretty downtrodden because, when I was out of earshot, the owner said to the realtor, "There's no way that guy can afford this place."

I walked around to the back and saw a nice-looking tractor. I decided it would be a good piece of equipment to include in the purchase. Marilyn liked the house and so did our daughters, so we were set on buying it. When we were ready to complete the deal, I took a second look at the acreage. When I arrived this time, the weather was a little better, so the owner and the realtor came with me. I especially wanted to take another look at that tractor. When I reached the spot out back, the tractor had mysteriously aged badly over the two days since I'd seen it, and it looked pretty beaten up.

I turned to the realtor and said, "I want the other tractor put back here." The owner hadn't thought I'd care either way, but within a day the tractors had been traded. When it came to pay up, we paid $20,000 in cash. We bought it in summer, but our Coquitlam house took until November to sell, so we had a couple of months to gradually move our things over to the Cloverdale house.

Right away I got to be friends with the guy next door. He had a hundred-acre farm beside our twenty and was leasing it to some young fellas who were running a dairy farm. He lived in Washington and was thinking he might put the property up for sale. In the meantime, I had bought twenty head of beef cattle, one for our food and nineteen for market, and I asked my neighbour if I could run my cattle into his field. He said sure and offered to sell me his hundred when he decided to let it go. The barn was in really bad shape and the young guys were struggling, so after a couple of years he stopped renting and sold off the dairy. A few months later I bought his acreage.

My other neighbour—Farmer Brown, I called him—owned three hundred acres of potatoes. After he died, his son Jack Brown took over and started running cattle. He came over one day and said, "Why don't you open up your hundred acres and I'll open my three hundred, and we can run our cattle together."

I said, "If you ever want to sell the three hundred, I'd be interested."

At the time he wasn't, but a short while later he came over again and said, "I've been thinking about it and I'm gonna put my farm up for sale next year."

"Let's talk again," I said, "when you get closer to the time."

After he left, Marilyn looked at me like she couldn't believe what she'd just heard. "What're we going to do with three hundred more acres?"

"I'm not sure. I only know that land is a good investment."

"But if we buy his three hundred and put it together with the other hundred and our twenty, we'll have four hundred and twenty acres! What will we do with all that land?"

I didn't really have a good explanation for her, but I knew it would be a good place to put our money. She trusted my judgment and we always managed, so she left me to make the decision, which I did a year later. I invited Jack Brown to come to the mill to discuss the price, and I got a fair deal in the end. So now I had 420 acres all connected.

When we lived in Coquitlam, I was a member of the Vancouver Golf Club. Pat Duffey was the manager, and one day he phoned me at the mill. I knew Pat to say hello but had never really had a conversation with him, but my accountant Neil Russell was also a member of the club and knew Pat well. Apparently, he had mentioned to Pat that I had a few sawmills in the valley.

I took Pat's call and, after polite chit-chat about the Vancouver Golf Club, he said, "I understand you've got some land."

"What're you thinking?"

"It's a big chunk of land just sitting there."

"You thinking about a golf course?"

"Maybe."

"Maybe?"

"Yeah, I'm thinking a golf course."

Part of our 420 acres were in the Agricultural Land Reserve (ALR), an area of land that covered about eighteen thousand square miles and had been set up by the New Democratic Party government in 1973 to encourage farming. In 1988, the Social Credit provincial government allowed golf courses to be developed on the ALR and, suddenly, Surrey had 181 proposals.

In June of '89, Pat Duffey submitted our proposal to the City of Surrey to get three hundred acres rezoned from agricultural to recreational land. Because of all the proposals going to the municipal council, our proposal had to wait while Surrey staff finished looking at the impact golf courses would have on the city. By November, the study came out showing that, as more people moved into Surrey, the city would need another nine to fifteen golf courses by the year 2000 to meet the demand. Unfortunately, ours wouldn't be one of them. The deciding vote came from an alderman who had once said, "Hell will freeze over before I vote in favour of sacrificing farmland

Arnold Palmer and I consider the possibilities for a golf course.

for golf." Council said it turned down our rezoning application because we hadn't met all the requirements.

The problem was we didn't know the rules when we first applied. Our original plan included about fifteen hundred yards stretching inside the ALR, and the mayor wouldn't support that big an encroachment. We went back to the drawing board and shifted the whole project to the south of the parcel, cut down the encroachment by 50 percent, and revised our proposal. The report had said we couldn't go more than 430 yards into the ALR, but we saw that as a guideline, not an absolute maximum. Pat and I figured the municipal study had underestimated the number of golfers in the area because the Surrey courses were always crowded, so we added into our proposal the guarantee that, 50 percent of the time, Surrey residents could reserve tee-off times more than forty-eight hours in advance.

While we waited to hear back from the city, Pat Duffey and I started talking about who we should get to design the course. The

two names we bounced back and forth were two of the greatest professional golfers in the game: Arnold Palmer and Jack Nicklaus. We
couldn't see much difference in the two, but we'd heard that Arnold
had built a course in Whistler and liked BC, so we went with him.
Pat and I phoned his office and a woman answered. We explained we
were thinking about building a golf course in our area and wanted
to know if Arnold would be interested in designing it. The woman
called back a little while later to tell us they would send a contact to
scout out the area and meet with us. The scout liked what he saw of
the acreage and told Arnold it was a good, workable piece of property.
Not long after that, Arnold arrived from Florida with a couple of his
men and we all walked the course.

His chief architect, Ed Seay, had been with Arnold since 1971 and
told me, "Arnold's one of the leaders in the design and construction
of outstanding golf courses."

"We know about the one in Whistler."

"That's one of about 141 courses throughout the States and inter-
nationally that he's been behind."

"Sounds like our guy."

The second time Arnold came out, I drove over to the Surrey
hotel in my '71 Chev pickup and he threw his clubs in the back. "Just
in case."

As we drove around the acreage, I wondered if he thought our
plan was possible. We went from section to section, and he'd say,
"This'll be a nice place for a hole." And, "You've got some natural features here we can work with. We can keep all those trees."

He was also impressed that the Serpentine River bordered the
property to the north. "An excellent water feature." I could see him
figuring it out in his head. He was a nice guy, real friendly, knew
what he wanted to do, and at the end of the day he went straight to
the point: "This'll work."

That day, I looked toward the north and saw the North Shore mountaintops shining in the sun. "What a view," I said. Right then, I had a name for the golf course: Northview. Arnold nodded. "Sounds good."

Later in the evening while we ate dinner, he explained why, with all his other business ventures, he took the time to design so many courses. "I truly enjoy designing fun and challenging golf courses. But you have to consider the area. All of the great courses blend in naturally with the existing environment. We respect and try to preserve the terrain on every course as much as we can and still stay within the framework of playability. Put simply, we don't force or design unnecessary grade work or features."

"How much will you be hands on?"

"I've got a couple of guys on my team that work with me on the design. I give them my feelings and thoughts on a particular hole and they fit it in. We work very well together."

"Ed Seay's been with you a long time," I said. "You both must see things the same." I was thinking how Vic Rempel and I shared the same point of view about S & R.

"Yeah, Ed's a great asset. His basic philosophy of golf course design is the same as mine. I believe in a traditional straightforward design that produces courses with lasting quality and that are exciting and enjoyable for all players."

I tried imagining playing a few rounds with the "King" himself on our course, and then I remembered my level of skill.

When we finished dinner, Arnold asked if he could see the mill. I was happy to take him over during the night shift when it wouldn't be as busy. I think he wanted to make sure I had the means to pay him. We arrived around 10:00, and I started to show him where the logs come into the mill when I saw one of my Japanese customers standing nearby. I called him over and said, "I'd like you to meet

Arnold Palmer." The Japanese guy freaked when he saw him. He kept saying, "It's Arnold Palmer! It's Arnold Palmer!" He couldn't believe he was meeting this golf celebrity in a sawmill in Surrey. Arnold laughed and shook the man's hand. I don't think our customer ever forgot that night.

Not long after his visit, we got the word that Arnold was good to go. We were glad to have him in our corner, especially when we returned to the City of Surrey to try again to get our land rezoned. Our enthusiasm was short-lived.

14

BATTLE OVER GOLF

N JANUARY OF 1990, Pat Duffey filed our new application. We included Arnold's design of a thirty-six-hole golf course and a 43,000-square-foot clubhouse. We pointed out that the property had safe and easy access to the site and that the natural borders would fit in with our farming neighbours. We also said we were going to put aside twenty-two acres for public walkways along the Serpentine River and canal. At the council meeting, one alderman said he was in favour of golf courses if they were built on land that couldn't be farmed, like land covered in bush and trees or sloped land, or land of poor soil quality for farming: "None of us would support golf courses on prime farmland, but if land is sitting idle a golf course is an excellent option." His vote was in the minority. Council voted 5–4 against our application.

The other three in favour saw golf courses as recreation for the community. They wanted to hold a public hearing to get wider public input, but those opposed wanted farmland preserved. The mayor was leading the fight against us: "If this goes through, the land in all agricultural areas will skyrocket so no legitimate farming interests will be able to acquire land. We'll run out of farmland in five years if this keeps up."

One alderman asked, "But Mayor, why not go to a public hearing?" "Because my mind's made up."

Pat Duffey was upset with Mayor Bose and the others who voted against us. He figured they were only thinking of themselves. "This was a political decision, Chick. They're not thinking about the next generation; they're thinking about the next election."

We had already set up slots for 148 jobs for local people to fill and could see that the course would benefit Surrey, but for now the door seemed shut tight. We were disappointed to see the plan defeated and decided to reapply in six months. But a couple of months later, we got a bit of luck and it looked like the door might open a crack.

Mayor Bose had gone to the newspaper about our proposal. He said we had misrepresented the soil quality in our application: "They've attempted to downplay the value of the land in soil class to make it look like poor soil." Pat shot back a reply that it was Surrey's own municipal planning department that classed the soil as low grade: "We've never done anything wrong. What Mayor Bose has done is wrong." The mayor denied any ulterior motives: "I know of the soil's potential in this area because I grew up at a farm here, and council's decision must be based on the land's potential." The council held a meeting and, while the mayor and two others were against it, the majority agreed to let us return in March with a revised plan.

In March, Pat and I went to the council meeting, hoping for the best. The mayor grilled Pat again about the class of soil, and Pat repeated that municipal staff had classified the soil at a low grade of 4, poorer than the good grades of 1 to 3. Pat held up a land assessment report and said, "The salt content of the soil would require a minimum of a million dollars to fix through chemicals and irrigation to make this agricultural land." He then reminded the mayor that members of the Bose family had leased a portion of our acreage, and they hadn't been able to make it a successful farm. "It's become a giant blackberry patch."

The mayor questioned the salt content of the soil again and said, "I don't believe you."

"Parts of your own family farm are the same," said Pat. "They're on a lower-class soil too."

"That is absolutely false. The Bose farm is viable and on high-quality soil."

When we left the meeting, we weren't hopeful. Pat shook his head. "I don't know how to deal with Bose's comments. He knows we're telling the truth, but I can see it won't do him any favours to admit he's got bad soil too. Who'd buy his land?"

I knew Pat was frustrated. We felt we were getting a raw deal, but I had to trust that the mayor was fair. "We'll keep trying," I said. "I believe it *can* be done."

In July, council refused to review our application.

"If he and the others don't budge on this," Pat said, "it could become part of the election in November."

Pat Duffey called it right! The municipal elections shifted the balance of power in our favour. The guy who had said hell would freeze over before he'd vote for golf courses was now out, and a man who said he would listen to both sides of the issue was on the council. That was the only change, but it meant a lot to our cause.

In January of 1991, an alderman who supported us asked council to reconsider our rezoning application, but the ones against wouldn't budge. They said we were trying to get this put through before the end of the year when a new provincial government might come in and stop all golf course development. They also wanted to keep the agricultural land because "we prefer cow-pies over golf balls!" The alderman continued to speak on our behalf, saying our site had terrible soil and saw no agricultural potential so

Northview's plan should go to a public hearing. Council voted to let our application go to the next stage in the planning process, and this time, the majority was in favour of giving us another chance. We went to work right away to present a strong proposal to convince council to approve our golf course.

Mayor Bose fought back. He went to the newspapers again and repeated that his family had farmed the area for a hundred years. "This golf course would be an astonishing abuse of farmland. I'm strongly opposed to this outrageous intrusion into the ALR. I'm very concerned about the implications this might have for Surrey's farmland."

Other aldermen told us that council still had the power to say no to a rush of golf courses. Not everyone would get permits approved. The council members on our side understood that most of the residents came from farming families and knew the importance of the ALR, but Northview's proposal seemed reasonable.

We were hoping that council would accept the report that showed most of our soil was useless for farming. The cost to make it workable would be a major investment. Why not let us use the land in a better way? We knew of another family applying for permission for an eighteen-hole golf course. They wanted to stay on the property and were hoping the change in zoning would keep them from financial ruin. They had a 138-acre potato field that had been wrecked by wire worms and made the land impossible to farm. The family had been trying to sell it for twelve years without any luck.

The other concern from the municipal council was for the wildlife, so we asked an expert to explain that golf courses can be set up to have land put aside for wildlife so golfers and birds can get along.

In May of 1991, we gathered supporters and reports to help us get the development permit approved. I attended the public hearing to speak to the Surrey council about the proposed rezoning. I pointed out that Marilyn and I were the sole owners of the property, and in response to their suggestion that we try farming I told them that there'd been six different farmers on my hundred acres over the years, and they had all left because they couldn't make any money. "Farming is not a viable use for this land," I said.

I explained that we were planning two first-class eighteen-hole golf courses, and that Arnold Palmer's organization was overseeing the project. At the time, we were going to have a public and private course, but later made both courses public because competitors were building golf courses in Surrey, and we knew we wouldn't be able to sign up enough private members to make a go of it.

At the council meeting we had a few supporters. Arnold Palmer sent Barbara Gonzalez from Florida to represent the Palmer Design Company, and she confirmed he was part of the project. The council had earlier asked what we were going to do about wildlife, so Donald Blood, a wildlife biologist, handed in a report showing how we were going to protect habitat and conserve wildlife by building a wildlife reserve on a forty-acre strip that wouldn't have turf. Mr. Cuff, the co-chairman of the Surrey Golf Course Association, stated that about three thousand golfers wanted a municipal golf course and pointed out that golf courses were quiet neighbours, they provided enjoyment for everybody, and they didn't pollute the environment. In short, a course would be an asset to the community. Another member of the association, John Seimens, put in a good word, as did Barry Tyrer from Trans-Pacific Trading, my good friend and our first link to the Japanese market. He told them that Surrey would lose money if golfers travelled to the United States to play and that a golf course was the best use for the acreage. A man from the Cloverdale Board of Trade made the point that Surrey would benefit from the increased tax base and tourism. He figured

Surrey had 190,000 golfers in 1990 but that number would increase to 223,000 by 1996. He said the courses would hire 150 people and the clubhouse would bring in banquet facilities and meeting rooms, which weren't available in the area. He ended by saying that I was a private developer and I was offering to build a municipal golf course at my own expense.

We also had people opposed to the project. The representative from the Naturalist Society worried that the golf course would spread into protected areas, and he wanted us to show what the environmental impact would be if we used pesticides. A number of people were opposed for the same reason: they wanted to protect the farmland and didn't want us to go forward until everyone was on board that the environmental impact would be small.

One of the aldermen said that we were dedicating forty acres of farmland as a wildlife refuge: "They're going to enhance this particular piece of property beyond what they already have. The only redeeming value that this farmland has now is that it floods every winter."

Mayor Bose took the floor as Mr. Bose, a private citizen. He said he didn't like the idea for all the reasons he'd given us before, and because he was raised on a farm across the street and his family was farming potatoes there. They were afraid the golf course would mean more people would be trespassing on their farm and using the dyke system. He didn't want people walking down farm roads and trespassing on private farmland.

Of the 143 people who spoke at the public hearing, he was one of thirty-three who said they were against the development. The other 110 speakers supported us along with the 5,500 people who had signed our petition, 97 percent of whom were from Surrey. Two years after we first went to Surrey council, we finally got our application approval!

However, in the fall of 1991, when the NDP government came in, it put a hold on the development. A year went by until, in June of '92, council brought forward Pat's request for the approval of a development permit. We hoped it would get passed because we wanted to get started on the project. We had some support from the council members, but again not enough. Mayor Bose, in particular, was still determined that the course wouldn't go through. Marilyn, Pat and I had gone to see him and he told us in clear terms, "You'll have your golf course over my dead body!"

Marilyn and I tried not to get discouraged, and Pat still felt pretty positive that we had enough support on council, but the mayor was giving us a hard time. He had thirty days to either sign the permit or ask for council to vote again. The newspapers in our area wondered why he hadn't signed in the first place. "I still have very serious concerns," he stated. They were mostly about us preserving a wetland habitat for winter birds, which we had said we were going to include in our design, but he wasn't convinced. People were calling the golf course "an ecological disaster" and "an eyesore" built "for the rich." He also said that council had approved the permit before the government had reviewed the impact the golf course would have on geese and ducks. We had promised to create forty acres for the wetland, but the wildlife agencies figured we needed another eighty acres above that.

Surrey council could still approve the permit without waiting for the Ministry of Environment to say yes. It'd have to be a pretty big impact for the minister to go over the heads of the council. Fortunately, Councillor Marvin Hunt saw the agencies' demands as unfair: "I don't see why the Stewarts should be held to ransom for something that's just a wish list. Those recommendations are a guideline. The federal or provincial politicians aren't willing to back them up with legislation because it's not appropriate. The wildlife people are asking us for things they have no authority to ask for."

The battle kept on for some time. The mayor and the conservationists were ready to fight as long as it took to save the farmland. They were angry with our municipal supporters: "The pro-development lobby group on council is breaking with council's longstanding directive to preserve prime farmland. They're carrying on with reckless abandon." I could understand the mayor's concern that our project might encourage more companies to dig up acres of land to develop, but I believed he should give some relief to owners who had low-quality farmland. The mayor was ignoring the people who said we had a shortage of courses in the area, and that the golf course would be a benefit to the community, especially from tourism; it would also support charitable organizations, be a place for children to learn and play golf, be a boost to the area, and be a beautiful natural setting for families to enjoy. Our supporters did their best to show the good side of the project.

My neighbour Jack Brown even went to the newspaper to speak his mind. He was the president of the BC Federation of Agriculture, and he figured people who wanted every piece of the ALR just for food production were off base: "Farming in the west end of the Fraser Valley is unnecessary because some of it's poor quality. Better to go into the fertile land east of Langley. If environmentalists and governments want to keep agriculture here, they need to loosen the rules around land development. Let the farmers sell the worst of their land at rezoned prices to let them put the money back into the farms. It's a nice philosophical approach to say all this farmland should stay in the ALR but it doesn't have any economic basis."

We read this and hoped it would move our opponents to give us a break. We didn't know if it did, but we got good news soon after the article came out. Council voted again, and finally it was 6–2 in favour of the permit. Mayor Bose had no choice; he had to sign. We were glad the fight was over, and we could finally bring Arnold Palmer to town. But the mayor had another plan.

He said things were not in order and needed more looking into. One of the aldermen fought back: "I think the mayor has become so involved in this issue that I don't believe he can emotionally separate himself enough to be impartial, and he's letting that cloud his judgment."

Mayor Bose went ahead and got together with a couple of environmental groups and two Surrey residents to block the approval. They asked the Supreme Court of BC to make the council's vote invalid. He pointed out that the NDP government had just repealed the 1988 Social Credit ruling that allowed golf courses on the ALR, so now all golf course plans were put on hold and ours should be blocked too. The mayor wanted the government to stop these applications and had said that he didn't want to allow developers to "ruin the agricultural land of BC." He had until the end of October to use a section in the Municipal Act to take back our application. The councillors supporting us said our application had received a final reading, and "final means final." They said, "If the mayor tries undoing the application, he'll have a huge fight on his hands."

What a nightmare! Pat and I were scratching our heads, trying to figure out what the heck would happen next. Then, several months later, we received good news. The judge rejected the petition that wanted to kill our project.

About four years had passed since that first application to rezone the land. If we had known how much administrative work it was going to take, I don't think we would have started down this road. But now that was over, and we were set to start construction.

One surprising turn happened a few months later, when we learned of another project up for rezoning. The mayor and his family were selling off 580 acres of sloping farmland that was just down the road from our planned golf course. They wanted to get the property zoning changed from suburban to urban, so they could start building townhouses. At the council meeting, the mayor left the room so the members could vote without him influencing them. The majority gave the go-ahead to his proposal to develop his land to urban densities, which meant he could build over four hundred condos next to his family home and barn. The council vote approving the plan made the mayor and his family multi-millionaires. I guess a selling point for new buyers would be that they'd be living next to a thirty-six-hole golf course!

A year later, by the summer of 1994, we were ready to open eighteen holes. We had spent $1.5 million on pre-design and development cost charges and figured the golf course cost close to $25 million and the clubhouse about $7 million. Our pro shop would soon be stocked, our Duffey's restaurant would have great Sunday brunches, and our Palmer Room could serve about eighty guests if they wanted fine dining.

We included a natural-grass driving range, chipping green, and a 24,000-square-foot putting green. The four nines were built in a square to allow courses to be played east–west or north–south. The western eighteen measures 7,191 yards from the back, enough to challenge touring pros when it gets windy; the 6,861-yard eastern eighteen is a split between trees and water; and the two courses—the Ridge and the Canal—graduate down to 5,200 yards from the front tees. The 563-yard par-five eighteen-hole on the Ridge course could break a lot of golfers. The experts said we'd built a high-quality course. Not bad for someone who's not even a golf course guy!

Opening day, our family and a few dozen friends got into golf carts and drove around the completed eighteen-hole course. Arnold

Marilyn and I greet Arnold Palmer on opening day after he arrived at Northview by helicopter from the Vancouver Airport.

had come up from the States to officially open the course by playing a round of golf with me.

The remaining eighteen holes opened the following spring. We were pretty happy to see the place come out even better than imagined. Marilyn and I had been involved in the planning from day one and had spent more time there than at the mill when Northview was in the building stage. Pat managed the place, and Marilyn split her time between Northview and the mill. We got the Vancouver Canucks hockey team to use it for their annual fundraising benefit for the children's charity Canuck Place. Our family also hosted a tournament for an Alzheimer's benefit. The hard work had paid off, and the golf course became a great addition to the area.

Even though I was an average golfer, I did get to play a round of golf again with Arnold. We were paired up with two sports guys— Marty Zlotnik and Pat Quinn—and my ball went into a sand trap.

Marilyn, Arnold and I celebrate the opening of Northview Golf Course.

I couldn't get the thing up and out of there. Arnie was my partner, so he figured he'd better help; he suddenly saw something overhead and pointed, "What's that in the tree?" The other two men looked up, which gave Arnie time to pick up the ball and throw it on the green.

A year later, I was over at Northview to meet friends in the Palmer Room for dinner. I saw Mayor Bose sitting at one of the tables. He saw me and came over to greet me. To his credit, he apologized for fighting us so hard. He admitted he was wrong: "I made a mistake; Northview's been good for the community." I accepted his apology.

The PGA tour was the biggest event we had at Northview. It put our golf course on the map. From 1996 to 1998 we hosted the Greater Vancouver Open, which was the first Vancouver tour in over thirty years. The Canuck Foundation—the charity of the Vancouver Canucks hockey team—was the beneficiary of the Open. From

1999 to 2002 we hosted the Air Canada Championship and watched Payne Stewart, Luke Donald, Graeme McDowell, Matt Kuchar, Adam Scott and Mike Weir play.

In September of 1999, more than forty thousand people watched Mike Weir win his first PGA event. This was also the first time a Canadian had won a PGA tour event in Canada in forty-five years. We were standing around the eighteenth green when the crowd started singing "O Canada." Mike told us he was in shock when he won: "That was an unbelievable day!" He made it unbelievable on the fourteenth hole, where he holed out from 159 yards with an 8-iron for an eagle two. At that point, he'd won the tournament.

For seven years some of the best golfers in the world came to Northview, looking to win the million-dollar prize money. We almost always had perfect weather and record crowds. Would we host another PGA tour? If we got corporate sponsorship and the great golfers wanted to come, we could do it; we've been ready since they left.

Now, we're up to about sixty-two thousand rounds played each year, and our two courses are the fourth busiest in BC. We have practice facilities and CPGA instructors. Our staff has created the Academy of Golf Instruction and added a regional fitness centre. We also have a coaching program for elite junior golfers. When we opened in 1994, the area had a few trees but was pretty bare. Twenty-two years later, our courses have beautiful gardens that include a wedding garden with ponds, waterfalls and a gazebo. A big change from the blackberry bushes.

In all the time we've been operating Northview, I never regretted asking Pat Duffey back in '89 to take our application to the City of Surrey. I remember asking him, "How long do you figure it'll take to get approved?" His answer: "At most? Probably a couple of months." In 2003 Pat died after a short illness. He knew he was sick and wasn't

Northview's Enchanted Garden is a beautiful spot for weddings and photos. COURTESY OF NORTHVIEW

going to be around too long, but what can you do? We were all sad to lose him and we missed him. Fortunately, he had taught Marilyn how to run the business, so she felt confident she could look after the golf course and keep up with her work at the mill.

At Northview, Marilyn became general manager and held weekly meetings with the staff. She oversaw everything and got along with everybody. She would also take her paperwork from the mill over to her Northview office and the Northview work to the mill.

The golf course became a family business. While our daughters Suzanne and Colleen were working at the mill, Wendy's daughter, our granddaughter Melissa, worked at Northview in the wedding and banquet department for several years starting in her early twenties. Wendy's boys, our grandsons Jason and Josh, started out at Northview on the grounds crew and have worked in the office over the past fifteen years. Jason was the assistant GM until May of 2016, when he moved to his own golf course on Vancouver Island, not far from Youbou, where I worked at my first mill job. Josh took

Northview Golf Course today. GARY CORP

over his brother's duties as assistant GM, so things have been running smoothly.

Northview became a place for us to catch up with the day's business. Marilyn and I would have dinner in the Palmer Room and I'd help her with the bills. She and I were similar in this way—we both liked to keep busy. Sometimes at night in the summer months, we'd go over to the golf course around 5:00 p.m. when everyone had gone, and we'd play eighteen holes before the sun set. If it was getting dark we'd just play nine holes. The mill was all work, but the golf course was a little of work and play.

You'd think after all the trouble we had getting Northview off the ground, we'd have had our fill of dealing with municipal councils, but we had to walk through those doors again several years later. This time Marilyn and I knew what to expect and were prepared. We were about to take on a new project that came out of a personal tragedy and was grounded deep in our hearts.

15

CZORNY
ALZHEIMER CENTRE

N 1982, at the age of sixty-nine, Marilyn's father Mike Czorny was diagnosed with Alzheimer's disease. Here was a guy who always remembered the names and sections of every rail line in his charge, but who was starting to wander from his home and forget where he lived.

Marilyn's parents lived a few miles away in Chilliwack, so Marilyn would drive to their house and stay overnight. At first, everything seemed normal. If I went along to visit, I'd find him out in the garden digging in his flowerbed, and we'd have a yak about fishing or farming. Mike always wore a pressed white buttoned-up shirt, sometimes with the sleeves rolled above his elbows. My mother-in-law Nancy said he wore a white shirt when he worked on the railroad laying rail ties with tar, and even wore it when he was shovelling manure into the wheelbarrow. The shirt would get sweaty but not dirty. Never figured out how he kept those shirts so clean.

He was a fisherman and hunter, a strong guy with a stocky build who liked to stay fit. He was a hard worker, happy-go-lucky, and even learned how to swim when he was forty-seven years old. He loved his family and his red Cutlass Supreme. One time I was talking with him about the car and his work, and I noticed he was having trouble finishing a thought or remembering the names of the guys he'd worked with over the years. This was unusual because he'd always had a good memory, but I didn't think much about it. Same thing when he'd get a little confused about a story he was telling, and he'd get mixed up about why he'd been telling it, or he'd quit halfway through, or he'd tell the same story over and over. The general feeling from all of us was that Grandpa was getting forgetful.

The realization that something more serious was going on came to me when we were out fishing steelhead at Spence's Bridge. Marilyn had made a point of telling me to make sure I got a nice room in a motel for him: she was starting to notice a change in his moods and wanted him to be comfortable. Unfortunately, the motel was full, so I had to make a bed in my van. I never told her we'd slept overnight in the van, but it was the best I could do and Mike didn't mind. He was pretty easygoing.

We got out on the boat, not expecting to catch anything, so the conversation was loose and easy. As I told him about one particular day fishing on the Babine, he'd listen and nod. Then he'd ask me a question about that day, and I'd answer him. A few minutes later he'd ask the same question. I said, "Mike, I just told you about that." I felt bad because I didn't want to hurt his feelings, but he didn't seem to mind. We sat in silence for a while until I felt a tug on the line and managed to catch a fish. We were both excited about our good luck. As we were heading back to the van, he asked me that same question again about my Babine story. I gave a short answer and wondered if his hearing was going. I couldn't figure out what was up.

I was about to understand when we got home. We had put the steelhead in a plastic bag in the back and I'd told him, "Don't say we got the fish. I want to surprise them." I figured my family would tease us about not being good fishermen and then we'd bring out the fish and have a laugh over it. When we parked in the driveway, Mike went to the back of the van and grabbed the fish. He walked into the house and the first words out of his mouth were, "We caught a fish." I knew then that something was wrong. In earlier days, he would have kept our secret. This incident stands out because it was the first time I noticed something was wrong and it was serious.

When he was still living in Chilliwack, he'd sometimes get part-time work. If the railroad was looking to hire a group of guys to fix some track between Chilliwack and Kamloops, they'd call him up to work, but they soon stopped hiring him. The mental decline was slow but noticeable.

One time, Marilyn's younger sister Charlene and her husband Don were visiting Mike and Nancy. Everything had been normal during the visit until they went to bed. Don woke up and saw a flashlight shining in his face. He heard Mike yell at him, "Who the hell are you and what're you doing in my house?" Took some time to calm Mike down. Charlene and Don had met in high school and been married for over ten years by then, so it was a terrible shock when Mike didn't recognize his own son-in-law.

The strange behaviour kept coming. Mike had been wandering outside away from the house, and Nancy would find him a few blocks away, lost, and not knowing how he'd got himself there. She'd be upset with him, but he didn't seem to mind. This wandering kept on for some time, and Nancy was getting scared that he might hurt himself, so we put alarms on the doors at their house. Nancy couldn't watch him every minute but now she could at least keep track of him. But the alarms failed to go off one night and she had no idea he was missing. Early in the morning, our sons-in-law had gone dirt-bike riding, and at 5:00 in the morning they saw an old man walking along the road. They suddenly recognized it was their grandpa, so they took him back home—ten miles away!

Nancy, Marilyn and I talked about his condition. We could see he'd been getting slower in his thinking, but his forgetting names and faces was a concern. Marilyn and Nancy took him to the doctor and were told his memory loss was a sign of normal aging. We already figured it must be, so we adjusted to the idea that it was just a bit of senility brought on by age. The shock came when we got a second opinion after another doctor ran some tests. Turned out Mike had Alzheimer's disease, the most common form of dementia. Now we understood why he never seemed to mind if you gave him heck: it wasn't sinking in to his head. Marilyn was upset for her mom and dad when she found out and even worried it might be hereditary because Mike's father John Czorny had Alzheimer's when he died in 1954. We talked about what we should do next and decided to move Mike and Nancy into the house beside ours when it became vacant.

Nancy was having a hard time of it. She was getting frustrated with her husband's moods. One minute he'd be talking normally to her and the next he'd start yelling because he thought she'd taken his cup of coffee away, but she hadn't done anything. The next minute he'd be quiet and sad. This was tough because he was usually a friendly guy with an easy smile, but now he just wasn't himself. He wasn't sleeping well, which kept Nancy awake. She told Marilyn she was even becoming too tired to do simple chores. Plus, he followed behind her everywhere and got in her way. "He asks me the same question and I keep answering it, but he forgets what I said. He even forgets he asked it, and then I get mad and he gets mad." She was embarrassed to admit she had yelled at him. "I just feel helpless."

Marilyn and I could see Mike losing his personality. Eventually, he didn't want to leave the house and didn't want to talk to anyone. He was leaning on Nancy for everything, and she worried she couldn't take care of him if he got worse. We tried to help her as much as we could, and Marilyn would try to cheer him up when he got depressed, but when Marilyn visited he would cry if she left the room. This illness was new for us and we didn't know where to turn, so we tried to do it on our own until his temper got out of control. He wasn't a mean guy; the disease was to blame.

At that point, we had to move him to Chilliwack hospital. Unfortunately, the place wasn't locked, so he kept walking off the ward and out the front door. He was a strong guy, so he'd just push people out of his way. Then the doctors moved him to a psychiatric facility and placed him in a white-walled room where the staff kept him in a straitjacket. He was there for six months, heavily sedated, and Marilyn couldn't stand it anymore, so we had him moved to Valleyview, on the grounds of Riverview Hospital in Coquitlam. This was a place for people with dementia, and he was well treated there, even starting to remember us again. But the sad fact was there weren't any proper places strictly for Alzheimer's patients to go.

Near the end, he stopped talking but would point to his Zane Grey cowboy books to get us to read to him. Three years after being diagnosed with the disease, he died in hospital in February 1985. He was seventy-two years old. Marilyn, her brother Sandy and I were at the hospital when he died.

This whole experience made us aware of the lack of proper care for Alzheimer's patients. Marilyn couldn't get the picture of him in a straitjacket out of her head. "That psychiatric ward wasn't a loving environment, and hospitals can't deal with dementia." Unfortunately, our choices at that time were pretty slim. Marilyn started thinking about what we could do to make things better for the patients. She began to put together an idea of a special centre that would be comfortable and homelike, and specially designed for the

Alzheimer's patients' needs, a place her dad would have loved. We got in touch with the Alzheimer Society of BC for help, and they were happy to work with us to get Marilyn's vision for the place off the ground. After that first meeting, Marilyn said, "We need to save our money so we can make this happen."

Over the next twenty years, we kept our sights on Marilyn's vision. In 1997, I noticed a for-sale sign on an 8.4-acre farm around the corner from our home, so I called our realtor and we bought it. We figured this would make an ideal spot for the centre.

A couple of years later, we lost another member of our family: Nancy Czorny died. Marilyn decided that when the centre was complete, she wanted to have a photo of her mom and dad inside to honour their memory.

A few years after that, we met with the Fraser Health Authority, an organization in our area that's in charge of getting good health care to over a million people. We knew it would be a sound partner. Together with the Alzheimer Society, we came up with a good design and then crossed our fingers as we went to the municipality to see if we could get the property rezoned.

We hoped the council would approve the rezoning faster than they had our golf course project. Fortunately, we knew Pat Higginbotham and his wife Judy, who was a councillor in Surrey. Judy said she would help us through the approval process. First, we went to the Agricultural Land Commission in February of 2003, and they approved the site for a non-farm use to develop phase one of our non-profit Alzheimer's facility. In 2004, council approved the property to be rezoned from residential to comprehensive development. This particular zoning let the municipality make sure we were environmentally sensitive throughout the project. In June of 2005, Surrey council approved our development permit that included a condition that we keep the one-storey rancher/farm-style character.

Breaking ground on the Czorny Alzheimer Centre. Left to right: Surrey councillor Judy Higginbotham, Bob Smith, CEO of the Fraser Health Authority 2002–2006, Loretta Solomon, executive director of the Surrey Memorial Hospital, Marilyn, me, and Rosemary Rawnsley and Dan Eisner from the Alzheimer Society of BC.

We had already started improving the soil on the acreage for the following year's planting. We complied with all the regulations and by-laws because we'd been through delays before and didn't want to throw a spanner into the works.

Council had approved phase one of the facility. The building would be a 34,102-square-foot structure with thirty-six rooms with single beds and ensuite bathrooms. The residents would share a common living room, dining room, kitchen and recreational and utility rooms, and we also put in a spa and a hairdressing salon. Outside we would build gardens and circular paths so no one would get lost. Marilyn wanted the whole place to feel like a home, and the gardens would give the people a place to be calm if they became confused or frustrated. We included extra space for sixteen to twenty adults in an adult daycare program, an Alzheimer Resource Centre, a dementia research facility, and administrative offices. Marilyn was more involved in the design because I was busy working at the mill, but we both worked together to keep the money in the account. We managed to donate $10 million in capital funding to get the place up and running.

In 2007, after twenty-two years of dreaming about it, we opened the Czorny Alzheimer Centre. Marilyn told the press, "Our prayer is that this centre will ensure that people with Alzheimer's and their families will receive the care, support and hope they need." We received a nice note from Canada's governor general, Michaëlle Jean, a patron of the Alzheimer Society of Canada: "I am impressed by the support shown by everyone who has devoted themselves to the success of the Czorny Alzheimer Centre; your labours do not go unnoticed. They are celebrated, respected, and admired by those whose lives you have touched."

During the opening-day event, BC and Surrey politicians thanked us for our contribution, and the people at Fraser Health said the centre could become a model for the country. "This is like a family home, not an institution, with communal kitchens and fireplaces, and the people who will reside here will be residents, not patients. They will not be 'admitted,' but will 'move in.' This will be their home for life. As the disease progresses, their care needs will change and the behaviours will require different skill levels, but we will have the best people working here 24/7."

Staff and volunteers were brought in to provide person-centred care in a gentle and peaceful environment. We got my doctor, Dr. Willms, to be the centre's resident physician. He'd been a doctor in Cloverdale a long time, and people saw him as a wonderful, humanistic physician.

TOP The entrance to the Czorny Alzheimer Centre.

BOTTOM One of the six cottage dining rooms.

The centre has a special design of three cottages with twelve residents living in each one. After going through Mike's experience in a psychiatric ward, we wanted to provide psychological security and comfort in the cottages. Marilyn named the cottages after

Marilyn, her brother Sandy and sister Charlene at Northview Golf Course in November 2012.

her favourite flowers: Rose Garden, Sweet Pea and Pansy Lane. She had worked hard to change the old institution model to a homelike centre. She told the guests on opening day, "I wanted to provide a nurturing environment that will enhance the lives of the residents by focusing on the abilities, strengths and interests of each of them. We picked this site to build a centre so I could walk across the street and hold their hands and take them for a walk in the garden. We put the money aside from the business and have worked like heck on it ever since."

Through a lease agreement with us, Fraser Health Authority would run the facility. They would have to hire registered nurses, psychiatric nurses, a music therapist, recreation therapist, social workers and rehabilitation staff. The design and staffing needs had been banged out over several years during meetings with Fraser Health and the Alzheimer Society of BC. At those meetings, Marilyn and I heard that various forms of dementia affect over 70,000 people in BC, and it's estimated that 937,000 Canadians will have the disease by 2031 if there's no cure found.

Our thirty-six beds were a start, but we were ready to move on to the second and final phase: adding another thirty-six rooms. This would mean building an additional 30,000-square-foot structure and getting more approval from the municipality.

In 2009, we returned to the Surrey council for approval of a report on our development permit application to expand Czorny. Fortunately, within a short time, we were ready to start building.

In June of 2012, Marilyn and I saw the completion of the final phase. The Czorny Alzheimer Centre now has seventy-two publicly subsidized residential care beds. The six cottages have twelve bedrooms in each one. The three new cottages also have flower names: Poppy Place, Lavender Way and Sunflower Garden. We included a coffee shop, an old-fashioned general store, a library and a quiet space to worship. Years before, I had bought a Model-T Ford off a buddy in Coquitlam, and it sat for a while in the lobby at Northview. We decided to move it over to Czorny and place it in a spot that looks like a garage, so the residents can tinker with it and polish it or sit in it and remember an earlier time. I also had an antique wood cookstove that we figured would bring back memories, so we donated it to the centre.

The Czorny centre has been a great success. We knew if we kept to Marilyn's vision, we could get it completed the way she wanted. Don Hickling is the vice-president of Leadership Giving at the Jim Pattison Outpatient Care and Surgery Centre, and he called Czorny "the largest gift to Alzheimer care in Canada," saying the final design would never have come from ordinary bureaucrats. We had to agree. At first, the professionals thought it was an impossible vision, but Marilyn had stayed firm on giving the place that special touch of love, and her determination paid off. Her brother Sandy and sister Charlene agreed their father would be proud.

Occasionally, after Mike was diagnosed with Alzheimer's, Marilyn would worry that she might one day inherit the disease. She didn't like to think about it, but her grandfather and father had had it, so the chance of her getting it was high. Turned out she *didn't* get the disease, but we were all saddened by the recent news that her younger sister Charlene has been diagnosed with early onset Alzheimer's. She came to visit and said she had known for a couple of years because she was forgetting things and her memory was bad. She and her husband Don had gone to the doctor and she'd taken a test—the doctor showed her a list of six things and asked her about them a few minutes later. She remembered only a couple of them.

Don took Charlene to the University of Calgary, where they got her into a study. The testers showed her a picture of a house and asked her to copy it, but she couldn't do it. "I knew something was really wrong," Don said. The CAT scan diagnosed the disease. They could see a shrinking of the brain. Marilyn's brother Sandy had just died in hospice care, so this news was difficult for the family. As we get older, we know we aren't going to live forever, but Charlene was in her early sixties, still young. And last year, my brothers Sam and Dave died within a couple of weeks of each other, which hit me pretty hard because several months earlier I suffered my greatest loss.

Charlene told me, "I'm forgetting some things, but I still remember when Marilyn phoned me and said, 'I've got some bad news for you.' I waited, and then she said, 'I've got cancer and I've got one or two weeks to live.'"

16

SAD GOODBYES

I TURNED EIGHTY-EIGHT in September of 2016, and I feel pretty good considering I was recently diagnosed with Parkinson's disease. This is the same disease that my older brother Dave had for twelve years, so I guess in some cases it can be hereditary. The slight tremors in my left hand got the tests started, and the doctors gave me medication to keep things under control. I can't expect to be in the same shape I was twenty or thirty years ago, but I'm still going into the office, and I swim every morning. Can't drive anymore, but my family have stepped in to take over the steering wheel, so I'm doing all right.

Like most people who get into their ninth decade, I've lost a few loved ones along the way. In 2012 my brother Denny, born a year after me, went first. He had been married for sixty years and had worked at the mill. A happy guy who lived a full life. Marilyn and I were on our way to Cabo, Mexico, and spoke to him before we left. He said he wasn't feeling well, and a few weeks later he passed away. His heart had given out. At Denny's funeral I kept thinking about the day we moved to Vancouver. Dave, Denny and I had walked several blocks to see *Sergeant York* at the movie theatre, and on that same day Dave and I got jobs as pin setters at a bowling alley on Broadway. We had to send Denny home to tell our mom we were working and would be home late. Denny was still pretty young at the time and a little teary-eyed when he said, "I don't know where we live!" He found his way home that evening, and later through life.

Our sister Margaret followed Denny. She had been married twice and found a good man the second time around. Her husband worked at S & R for a number of years before retiring from there. They were part of our family group that gathered in Haney at a Chinese restaurant every special birthday. Always a good time with plenty of memories to share.

Last year Sam, the third youngest, was in hospital in Kamloops. We were on our way there when we stopped in at Pitt Meadows to see my older brother Dave. His care worker of over a decade was crying when she told us Dave had passed away early that morning. We were stunned as we drove to Kamloops. We worried that Sam might be in bad shape and weren't sure if we should break the news to him. He was alert and able to talk, so we ended up telling him about Dave. Within a couple of weeks, he was moved into a hospice. We continued to visit him there for the next two and a half weeks. The hospice was a nice place with single, private rooms and a comfortable couch in each of them for visitors. The nurses were wonderful to him, and the peaceful and quiet atmosphere helped us say our goodbyes. I never knew they had places like this for terminally ill patients. Sam died twenty days after Dave.

My youngest brother Herb and I are the only Stewart kids still kicking. When the two of us were at Dave's funeral, we saw a woman Dave had wanted to marry when he was a young guy. A Haney girl from years past. The woman's mother was against the marriage because we boys had reputations for being a bit wild and people figured we wouldn't get anywhere. The mother had told her daughter to put Dave out of her head because, as far as she could see, "Those Stewart boys won't amount to anything." Didn't bother me; our own parents thought we turned out okay.

The losses have been mounting these past few years and I've endured a few knocks, but the worst was when Marilyn died on October 24, 2014. I should have been the first to go because I'm eight years older, but fate opened up a hole and I fell in. These months without her have been the loneliest of my life. Writing this is really difficult.

Five weeks before she died, we were helping open up the Babine camp just like we'd done every September for years. I noticed she didn't feel well, but she never complained about anything, so I let it go. She had been going to different doctors for pains in her knee and for being sick on and off, and they had given her medication, but they couldn't figure out why she was in so much pain. Her doctor said she had to go to Surrey Memorial Hospital for tests. The night before she went in, she got her hair done: if she was going to stay in the hospital for a couple of days, she wanted to look nice. Plus, it perked her up.

After three days, Marilyn was not improving. We realized she would be in hospital a while longer, so we made sure she was never without a family member beside her. We were worried but tried to keep positive when they wheeled her down for more tests. Maybe they'd figure out what was wrong and help her get through it. She had stopped eating but was drinking water, and we kept thinking she'd pull through. But soon, she was in and out of consciousness, and a terrible sense of dread came over all of us.

She had been in the hospital for three weeks. Those last few days were hell when we knew she wasn't coming out. She had been fighting hard to keep alive. She never wanted to give up, but she just couldn't go on.

The hospital staff had originally put Marilyn in a two-bed room. The patient in the other bed had visitors day and night, so the room was noisy, not very peaceful. I talked to the doctor and asked if we could have a private room. He said he'd arrange it for us. That day they moved Marilyn into a single room in palliative care. We were able to visit her by ourselves without strangers in the room. I saw then how important a quiet environment is when a family gathers for their loved one's final days.

Soon after the move to the private room, the doctor brought my daughters and me together and said there wasn't anything more they could do. They could keep Marilyn on life support if that was what we wanted, but she wasn't going to make it. I told my daughters we had to let her go. "Mom would never want to be kept on life support." We stayed with her until she passed.

The final test from the doctors came back a short time later—too late for Marilyn to learn which type of cancer she was suffering from. She never knew that the pain and fatigue had been caused by non-Hodgkin's lymphoma.

We had been married for fifty-eight years. We never fought and always discussed everything together and made decisions together: we were partners at work and home. We raised three beautiful daughters who gave us nine grandchildren and (so far) fourteen great-grandchildren.

Marilyn was a wonderful role model who always put others first. She believed she was on this earth to help and love people, so her plan for the Czorny Alzheimer Centre gave her another opportunity to care for people. She worked hard in every part of her life: at the mill, at Babine, at Northview, at Czorny and at home. She always said we were a busy couple and believed our good luck was the result of hard work. Together we continued to work hard, so we could enjoy

The Chanasyk family photo was taken in August 2016 at March Meadows Golf Club in Honeymoon Bay, BC. Left to right: Carter Chanasyk, Laura Chanasyk, Travis Chanasyk, my grandson Jason Chanasyk, Barry and Wendy Chanasyk (standing), Tawnni and my grandson Josh Chanasyk (seated), Chad de Groot, Easton de Groot, Blayke de Groot and my granddaughter Melissa de Groot. LONI SEARL OF TUESDAY PHOTOGRAPHY.

our lives, whether it was fishing on the Babine, sitting by the fireside at Gossip Island, playing nine holes of golf in Palm Springs before sunset, or relaxing in Cabo. We shared a strong work ethic and a love for our time off.

In fall of 2015, my daughters and I received more bad news from Kamloops: Marilyn's brother Sandy was dying. We drove up on several weekends to visit him and watched as he was moved from hospital to hospice. Now, sadly, my sister-in-law June is suffering a loss much like mine.

I am continually surprised that the weight of these deaths hasn't put me in the ground. How do I and every other person who has suffered loss survive? I remember saying to one of my daughters,

TOP The Dahl family in April 2017 at my daughter Suzanne's house in Langley, BC. Back row, left to right: James Lory holding my great-grandson Jonathan Lory, my granddaughter Jaclyn Lory, my daughter Suzanne, my grandson Jeffrey Dahl, Megan Vanderhoek (Jeffrey's girlfriend), Kevin Storrier; in front of Kevin is my granddaughter Michelle Storrier holding my great-granddaughter Paige Storrier. Front row, left to right: my great-granddaughters Hannah Lory and Haley Lory, great-grandson Lucas Lory, great-granddaughter Emma Storrier and great-grandson Kane Storrier. MELISSA DE GROOT

BOTTOM The Pollon family in June 2016. Left to right: Amy, wife of my grandson David, who's standing beside her, grandson Jeremy, Colleen's husband Dave, Colleen and granddaughter Jennifer.

"I don't know how we're going to get through this." And she said, "I don't either. But we will." I'm learning that as days go on, the deep hurt passes and becomes something back behind me, further away, but not so far that I don't still catch myself seeing Marilyn in the kitchen putting on the coffee while I crack eggs into the frying pan. Breakfast was a morning ritual we did together for years. Or we're sitting at the Babine having a glass of wine after the day's fishing. Or grocery shopping for the weekend late Friday afternoon, and then scooting across in the boat from Crescent Beach in time for an evening of cribbage at the cabin on Gossip. Or practising on the three-hole pitch-and-putt in our backyard. Or listening to the radio and a Glenn Miller big-band song like "Little Brown Jug" or "In the Mood" starts playing, and there we are dancing together again—a perfect fit. Marilyn has gone, but images of her appear unexpectedly and remind me how lucky I was that she said, "Yes, I'll marry you."

What can I do now? Turn away from life or put my energy to some good? One of the strongest impressions I was left with in Kamloops was the serenity of the hospice where Sam and Sandy spent their last days. The idea that people could die quietly in a room, without other people tracking all over the place and making noise, and be in a peaceful environment surrounded by family inspired me to put together a plan for a hospice in memory of Marilyn. I spoke with my daughters as well as the group that built the Czorny Centre, and we decided the perfect place would be on three acres of my property. The hospice in Kamloops has a beautiful view of the valley, so the view would be an important part of the design. We want our hospice to overlook the golf course and be situated where the family and the patient will have a view of the mountains in the distance.

We spent several days visiting different hospices in the Lower Mainland to see what we liked about each one. We want the design to include adequate parking, a visiting room and family room that

Here, Marilyn and I are in our final photo together, taken on my 86th birthday in 2014.

are large enough for a number of visitors, and areas that are big enough for people to get up and walk around. We noticed from our trips to other hospices that the entrance shouldn't be crowded and the kitchen and dining areas need to be accessible if a friend drops by for a visit. We want home-cooked meals made on the premises as well. All of these factors are being included in our design. We also studied different styles of buildings. We preferred two separate areas adding up to 24,000 square feet, with ten beds in each area for a total of twenty beds. We're thinking of calling it the Stewart Family Hospice.

We're in the planning stages, and I look forward to participating in each phase. This project allows me to keep involved and to stay busy. We estimate the project should be completed in less than two years, so for the next little while my life is pretty much mapped out.

The hospice, the Czorny centre, and our other projects would never have been possible without the support of the people who helped me throughout my life. All the employees at S & R, the men

I'm surrounded by my daughters Suzanne, Wendy and Colleen as we stand on the section of land designated for the Stewart Family Hospice. CLAUDETTE CARRACEDO.

and women on the Babine and at Northview, the customers who remained loyal since those early days back in 1963 when Vic Rempel and I took a chance and bought our first mill. Those people deserve a big thank you. I want them to know I appreciate every one of them and each of their contributions to my life.

As I complete this final chapter of my book, the past eighty-odd years seem to have lasted a fraction of time. I've suddenly become old. I'm surprised to see myself this way, but the stronger thought is gratitude that I was blessed with a loving partner and loving family and friends. I don't expect to live forever. I'm lucky to still be here, so for now, I'll go into the office, have lunch with the guys, and live until I'm done.

ACKNOWLEDGEMENTS

WOULD LIKE TO thank my daughters Wendy, Suzanne and Colleen for their help with my story. They encouraged me to keep at it and took time away from their own work and families to read chapters, offer comments and choose photos. Their love and support made this project even more enjoyable for me.

Thank you to Don Hickling who suggested I write this book and to Jackie Lee-Son who introduced us to Roxanne Davies and Michele Carter. Roxanne did a great job on organizing boxes of old photos that still hold wonderful memories. A very special thanks to Michele who put in countless hours making this book happen. It was truly a pleasure working with Roxanne and Michele, and I will never forget drinking tea and laughing while we turned the pages of my life. I enjoyed our time together.

Also many thanks to Alma Lee and all the folks at Harbour Publishing who made this book a reality.

I reflect on all who have helped me complete my life story. This is a book of memory and my memories have aged with me. I've made every effort to ensure that my recollections and research are accurate. I hope I got it right. A sincere thank you to my family, friends and employees. I couldn't have done this without you all.